T.T. BROWN'S

ELECTROGRAVITICS

RESEARCH

By

Thomas Valone, PhD, PE

INCLUDES:
"T.T. Brown Experiment Replicated"
reprinted with permission from
Electric Spacecraft Journal,
Issue 14, 1995

INTEGRITY RESEARCH INSTITUTE
5020 Sunnyside Ave. Suite 209
Beltsville MD 20705

To Thomas Townsend Brown who saw the future and went for it

Acknowledgement given to Mark McCandlish who is a modern day electrogravitics hero in spite of personal risk

T. T. Brown Electrogravitics Research edited by Thomas Valone, PhD

Fifth Edition......................2014

ISBN 978-1-935023-27-2

Report #603

Published by 501(c)3 nonprofit organization

Integrity Research Institute
5020 Sunnyside Ave., Suite 209
Beltsville MD 20705
800-295-7674
301-220-0440

www.IntegrityResearchInstitute.org

IRI@starpower.net

T.T. Brown Electrogravitics Research

TABLE OF CONTENTS

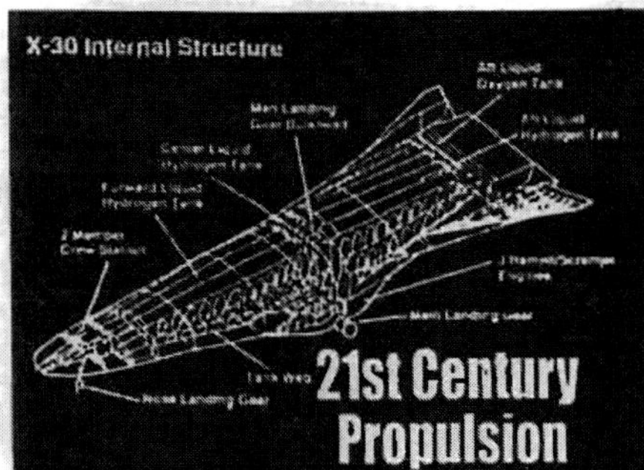

Electrogravitics

Research in the 1950's

Thomas Valone, MA, PE
Integrity Research Institute
1422 K Street NW, Ste 204 • Washington, DC 20005

Introduction

Electrogravitics, sometimes also called "gravitoelectrics", is the science of using high voltage electricity to provide propulsive force to aircraft or spacecraft of certain geometries. Its discovery is credited to Thomas Townsend Brown, a physicist who learned the secret from his professor, Dr. Paul Biefield. (There are indications that Biefield/ Brown experiments were predated in 1918 by Nipher's[1])

Unknown to many non-conventional propulsion experts, T.T. Brown's electrogravitics work after the war involved a classified multinational project. American companies such as Douglas, Glenn Martin, General Electric, Bell, and Sperry-Rand participated in the research effort. Britain, France, Sweden, Canada, and Germany also had concurrent projects from 1954 through 1956.

This article summarizes the recently declassified 1956 military document, "Electrogravitics Systems" by the Gravity Research Group of London (Special Weapons Study Unit)[2]. We will review the research highlights, the power range, electrogravitics theory, counterbary and baricentric control. This will show that Brown's discovery was a vital new addition to the aviation industry that became an integral part of the B-2 Stealth Bomber today—giving it an unlimited range.

Initial Studies

In 1929, T.T. Brown published his now famous article, "How I Control Gravity", which outlined his experiments and principles of his electro-gravitation theory.

A curious fact revealed in this first article is the alignment of the *molecular gravitors*. These massive dielectrics provided the most propulsive force when the "differently charged elements" were aligned (with the voltage source). This sounds like crystal plane alignment and perhaps explains the article "Gravity Nullified: Quartz Crystals Charged by High Frequency Currents Lose Their Weight" which appeared in the same magazine but was rescinded by the editors shortly afterwards.

T.T. Brown's first patent, US Patent #1,974,483 issued in 1934, was entitled "Electrostatic Motor" and is a fascinating free energy machine as well as a propulsion source. Claiming an efficiency of a "million to one", Brown causes the massive dielectrics to be the workhorse of the motor, exceeding, in his words,

> *the well known pin wheel effect or reaction from a high voltage point discharge.*

Much of what we know about T.T. Brown is from his numerous patents.

Though I was fortunate enough to correspond with him in 1981 when he was at the University of Florida. A sample of his detailed correspondence is contained in **Ether Technology** (which is a great introduction to Brown's work). An original drawing (shown below) of his shows the counterweighted saucer that optimized the charge distribution. The important detail is the power supply which went up to 250 kV, with a substantial force being displayed at 150 kV DC. Here we get an idea of the range of voltage necessary for successful electrogravitics.

Electrogravitics Systems

Only a few years ago, Dr. Paul LaViolette was in the Library of

Brown Comments on His Findings...

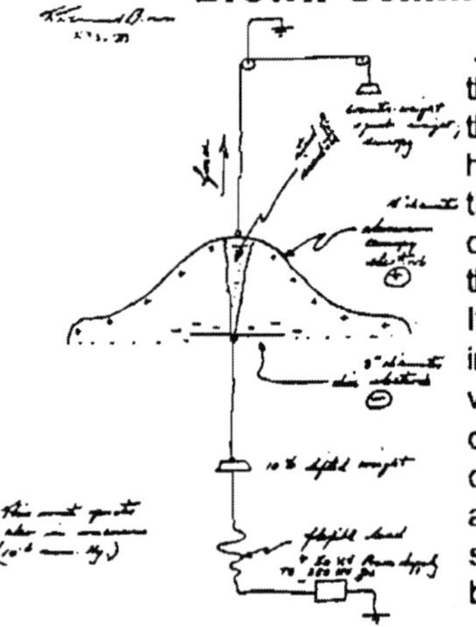

"... In reviewing my letter of April 5th I notice, in the drawings which I attached, that I specified the power supply to be 50 KV. Actually, I should have indicated that it was 50 to 250 KV DC for the reason that the experiments were conducted throughout that entire range. The higher the voltage, the greater was the force observed. It appeared that, in these rough tests, that the increase in force was approximately linear with voltage. In vacuum the same test was carried on with a canopy electrode approximately 6" in diameter, with substantial force being displayed at 150 KV DC. I have a short strip of movie film showing this motion within the vacuum chamber as the potential is applied."

CREDIT: *Ether Technology by Rho Sigma*

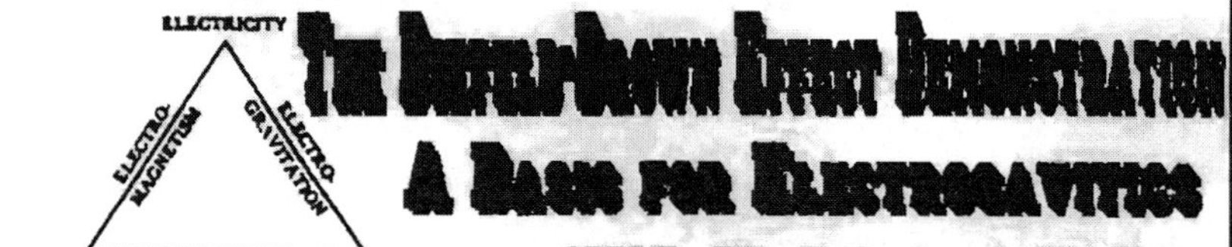

CREDIT: ***Ether Technology by Rho Sigma***

The first empirical experiments by Townsend Brown had the characteristic simplicity which has marked most other great scientific advancements, and concerned the behavior of a condenser when charged with electricity.

The startling revelation was that if placed in free suspension with the poles horizontal, the condenser, when charged, exhibited a forward thrust toward the positive pole! A reversal of polarity caused a reversal of the direction of thrust.

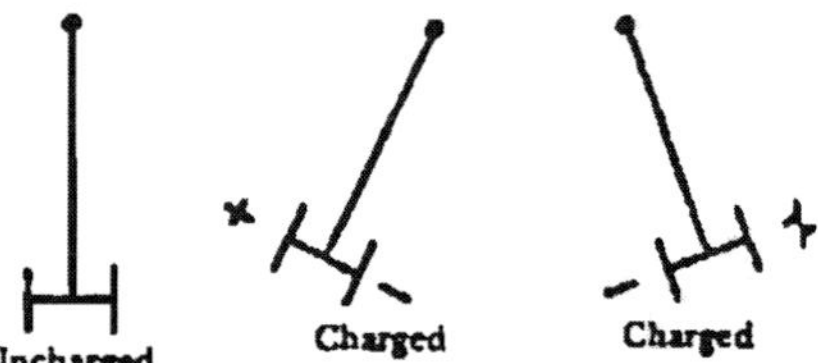

Further development of the implications of this phenomenon illustrated an "antigravity" effect. When balanced on a beam balance, and then charged, the condenser moves. If the positive pole is up, the condenser moves up (i.e., becomes "lighter"); if the positive pole is pointed down, it moves down [becomes "heavier").

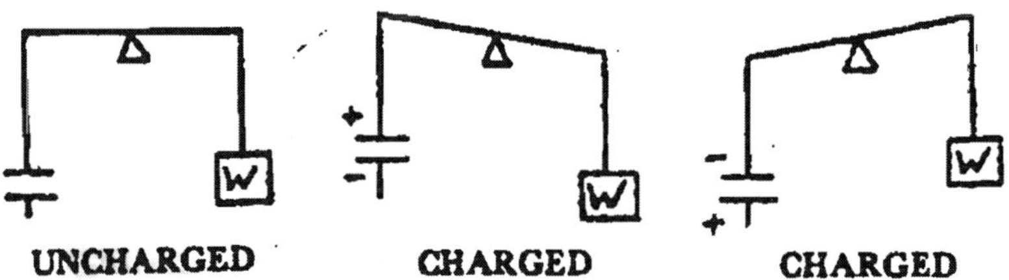

These two simple experiments demonstrate what is now known as the **Biefeld-Brown Effect.** To the best of our knowledge, it is the first method known of affecting a gravitational field by electrical means, and may contain the seeds of the control of gravity by Mankind.

The intensity of the effect is determined by five factors.

1. The separation of the plates of the condenser—closer plates, greater effect.
2. The higher the "K" factor, the greater the effect. ("K" is a measure of the ability of a material to store electric energy in the form of elastic stress.)
3. The greater the area of the condenser plates, the greater the effect.
4. The greater the voltage (potential) difference between the plates, the greater the effect.
5. The greater the mass of the material between the plates (dielectric), the greater the effect.

It is this last point which is inexplicable from the electromagnetic viewpoint, and which provides the connection with gravitation.

Use of Biefeld-Brown Effect can be used for electrogravitational steering of flying discs.

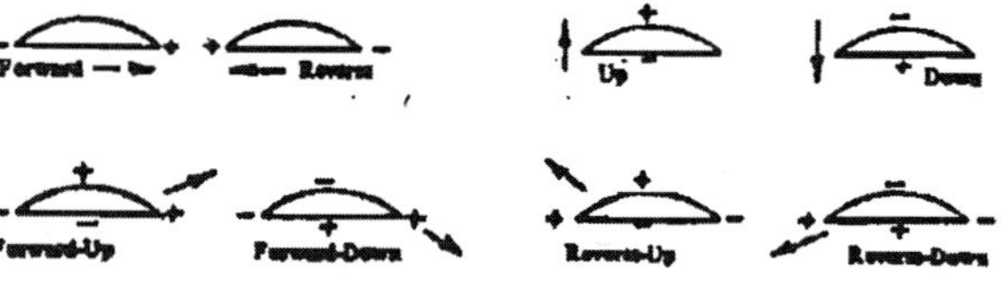

Congress in Washington, DC and looked up the work "electric" in the card catalog. Surprisingly, he found the listing for "Electrogravitics Systems"... a report that was missing from the stacks. When the librarian tried to locate any other copies, she commented, "It must be an exotic document" because she could find only one in the country which was at Wright-Patterson AFB.

Shortly after obtaining a copy, Paul and I agreed about the value of this report:

1) It validates T.T. Brown's experiments;
2) It lists the major corporations that were collaborating on electrogravitics;
3) The report expands the range of influence that Brown's research was having in the 1950's;
4) It includes the voltage and apparent power (KVA) requirements not real power (KW) for supersonic speed;
5) The report details the dielectric constant (k) goals for a Mach 3 fighter.
6) It shows the continuity from *Project Winterhaven* in 1952;
7) The report includes a list of electrostatic patents;
8) It had been *classified* by the Air Force for an undetermined amount of time which underscores its importance.

Prepared by the *Aviation Studies (International) Ltd., Gravity Research Group, Special Weapons Study Unit* in England in February of 1956, it defines electrogravitics as "a synthesis of electrostatic energy use for propulsion."

The report also asserts that:

> *electrogravitics had its birth after the War, when Townsend Brown sought to improve on the various proposals that then existed for electrostatic motors sufficiently to produce some visible manifestation of sustained motion.*

As mentioned in the first section labeled "Discussion", both *Project*

Winterhaven (1952) and *Electrogravitics Systems* (1956) postulate "a saucer as the basis of a possible interceptor with Mach 3 capability." Furthermore, a dielectric constant of 30,000 is presented as sufficient for supersonic speed (barium titanium oxide has 6,000).

This section also discusses the creation of a local gravitational system by the craft which "would confer upon the fighter the sharp-edged changes of direction typical of motion in space." The ironic part of this comment is that there was no space program then so it is excusable that they would believe a vacuum conferred sharp-edged changes of direction. The concept of a "local gravitational field" will be referred to later in this article.

Later in the report, we read:

> *One of the difficulties in 1954 and 1955 was to get aviation to take electrogravitics seriously.*

However, corporations such as Douglas, Sperry, Bell, GE, Hiller, Lear, and Convair are then described with an ongoing-project perspective. For example we read that,

> ***General Electric** is working on the use of electronic rigs to make adjustments to gravity.*
>
> ***Glenn Martin** say gravity control could be achieved in six years....*
>
> ***Clarke Electronics** state they have a rig, and add that in their view the source of gravity's force will be understood sooner than some people think.*

This information makes the report exciting reading and gives it an air of suspense.

Equally as impressive is the listing of M.I.T., CalTech, Princeton, University of N. Carolina, and the unique Gravity Research Foundation as "all active in gravity." The report goes on to talk about a Swedish company, two Canadian companies and a German company that have "woken up to the possibilities." In my view, the gem of the report is the admission:

> *Townsend Brown in electrogravitics is the equivalent of Frank Whittle in gas turbines.*

Though we may not know who Frank Whittle is, the comment makes T.T. Brown legendary.

The end of the "Discussion" section centers on the thought that theoretical breakthroughs to discover the sources of gravity "will be made by the most advanced intellects using the most advanced research tools." The physics which these 1950's scientists thought would be required to produce a breakthrough in the understanding of gravity centered around nuclear research and particle accelerators. They emphasize the "urgency of the matter" and look to the country with "the biggest tools of this kind."

In the "Conclusions" section, they also mention that a new bevatron may be able to "demonstrate limited gravity control", not really understanding that it is a field effect which controls electrogravitics. The definitions of two useful terms are made:

> **Counterbary** is the manipulation of gravitational force lines
>
> **Barycentric control** is the adjustment to such manipulative capability to produce a stable type of motion suitable for transportation.

The conclusion that a Mach 3 fighter is possible with "megavolt energies and a *k* of over 10,000" is a valuable piece of data in this section.

Dielectric Constant and Electric Field

Before going on to the Appendix of the first report, it is important to discuss the meaning of "dielectric constant." Defined as "farads per meter", the dielectric constant, *k*, is the same as "capacitance per distance", which is the capacity to hold charge and avoiding discharge or dielectric breakdown.

Capacitors, or charge buckets, are rated in *capacitance* (C) and also by *voltage* (V) limitations. Picturing a capacitor as two parallel plates, the dielectric is placed in between them, giving it more charge at the same voltage.

The equation for a capacitor is:

$$C = "ke_{o}A \bullet d$$

where

e_o is the permittivity of free space,
A is the area of the capacitor plate,
d is the distance between the plates.

This equation demonstrates that we can multiply the capacitor's ability to hold charge by its *k* value. Defining the dielectric constant of air as 1, we see that a *k* value of 10,000 would give a capacitor 10,000 times more charge in the same space limitations, which is important for aviation and for electrogravitics. The dielectric constant also gives the designer a direct control over what voltage he can use on the capacitor without experiencing electric discharges since,

$$Q = C\,V$$

where

Q is the amount of charge in coulombs

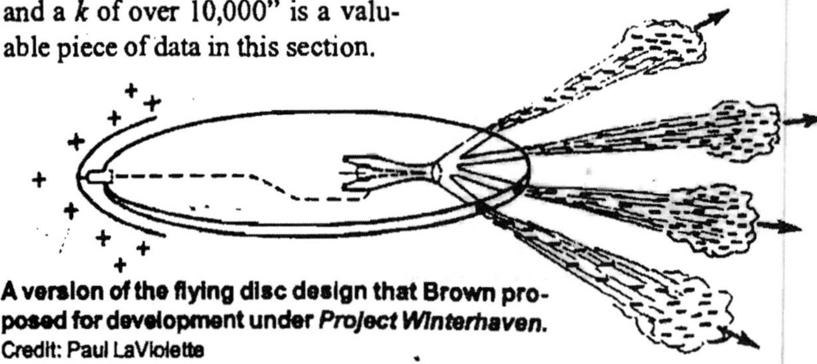

A version of the flying disc design that Brown proposed for development under *Project Winterhaven*.
Credit: Paul LaViolette

7

The point of mentioning this is that the charge in a capacitor is directly related to the electric field by an integral relationship. For a parallel plate capacitor, this becomes,

$$e_o kEA = Q$$

where

E is the electric field in volts per meter.

This leads to my primary reference which is the *Scientific American* article (Dec. '79) entitled "The Decay of the Vacuum" which describes the creation of matter and antimatter *without any input of energy* in the presence of an intense electric field. In what is called a *charged vacuum*, we should observe positron emission when an electron is accelerated through twice its rest mass, or about 1 Megavolt, in a short distance the size of an atom. The spontaneous creation of matter under such conditions is described by the authors as unstable empty space.

It is an intriguing fact that a supermassive nucleus, of atomic number Z=173 is sufficient to create this intense electric field and lower the total energy of the space. Creating a negative energy environment for the electron, the electric field would then be approximately 10^6V/10^{-11}m (where 1010^{-11}m is the Bohr radius) or about 10^{17} volts per meter.

John Searl's Device

It is worth mentioning in this context the work of John R.R. Searl, formerly of England, where he constructed numerous craft purported to fly with high voltage. Throughout the sixties and the seventies, J.R.R. Searl produced many newsletters detailing the work he was doing. Since I corresponded with him in 1981, I also received some of these reports.

The im ·rtance of his experiments lies in the phenomena associated with them. In the 6/1/68 issue of the *Searl National Space Research Consortium* newsletter, Barrett reports that the ionization of the air and permanent electric polarity of dielectrics were common along with the antigravity effects. In the 6/14/71 issue of the newsletter, Bernhard Vaegs reports that "a pink halo surrounded the craft" and "potentials on the order of 10^{13} to 10^{14} volts" were generated. This is consistently found throughout the reports and was measured by the length of discharge and the standard breakdown of air (10,000 V/m).

Barrett describes in the 6/1/68 issue the *electrets* that are used in the craft as well as a vacuum layer that surrounds the craft preventing ionization breakdown of the air. (*Electrets* are materials that hold an electric field similar to the way magnets manifest magnetism). Notice how close Searl gets to the 10^{17} V/m instability threshold of the vacuum.

The similarities between Searl's high voltage propulsion and T.T. Brown's high voltage propulsion are so striking—just the voltage generation

John R.R.Searl in England claims to have developed an anti-gravity device in 1949. Searl was employed by the Midlands Electricity Board as an electronic and electrical fitter.

The principle of it is allegedly that, when a metal annulus is rotated at sufficient speed, the conduction electrons are displaced outwards by centrifugal force, so producing a very intense negative charge on the outside perimeter and a positive charge on the inside.

method is different. Searl uses an unbalanced high voltage charge distribution to direct his crafts just as Brown does. We must conclude that the Searl propulsion method is *electrogravitic* in the same sense as T.T. Brown's method.

The Gravitics Situation

This is another major report that is included in the *Electrogravitics* anthology[2] was released by Gravity Rand Ltd. (div. of Aviation Studies) in December, 1956. It starts out with a quote from Einstein looking for a precise mechanism of gravity and gives an overview of new details such as the operation of Brown's saucers in a vacuum and management notes on the anti-gravity industry.

A fascinating glossary is included which defines such new terms as:

gravithermal, lofting, counterbary, baricentric control, negamass, magnetogravitics, etc.

A two-page journal reference list accompanies the report, with a summary of T. Townsend Brown's original patent specification.

Electrogravitics Voltage Disclosure

In the Appendix of the report (*Electrogravitics Systems*) we find a curious disclosure: "how, with the Mach 3 weapon in mind, to handle 50,000 KVA within the envelope of a thin pancake of 35 feet in diameter...." Since it is repeated later in the Appendix as well, it is important to distinguish this from 50,000 KW which is *real power*.

For example, electric utility bills are calculated based on kilowatt-hours (KWH) which is resistive load usage. KVA, or kilo-volt-amperes, describes *apparent power*, which is only used when AC power is in the system, such as with the volt-ampere rating of motors[7]. (The third part of the power triangle is *reactive power* which is power due to capacitors or inductors only.)

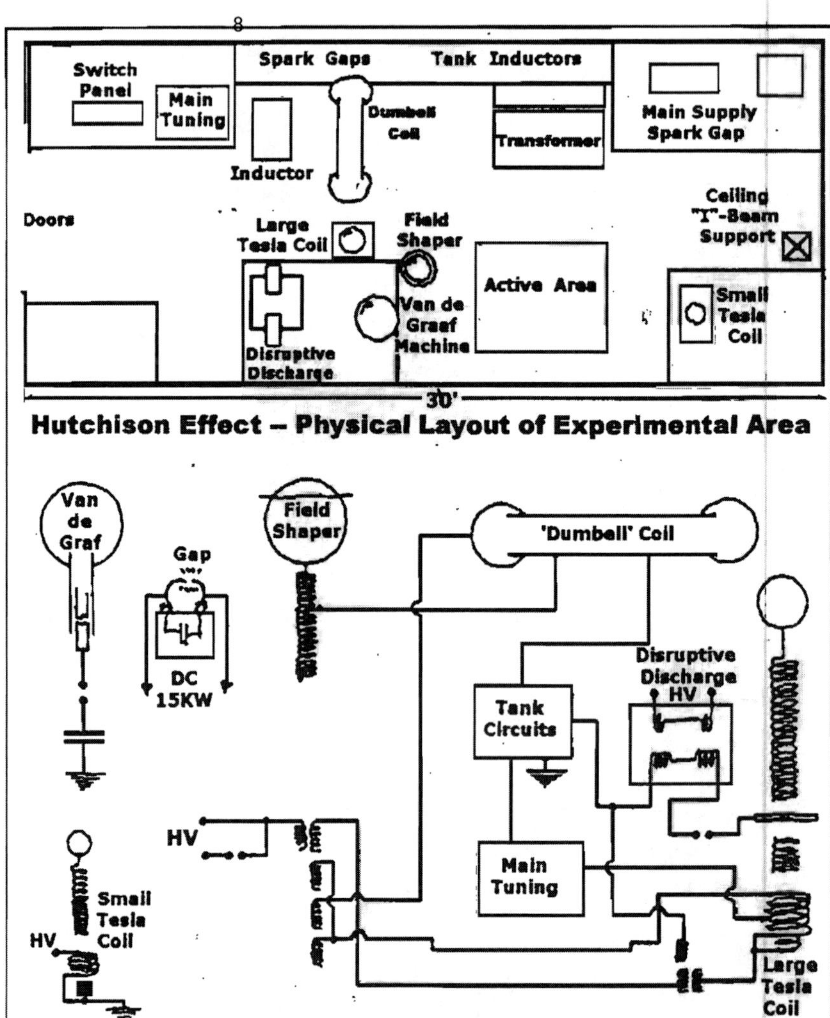

Hutchison Effect – Physical Layout of Experimental Area

Hutchison Effect – Circuit Block Diagram

It is known that Searl produces AC power but until the release of the *Electrogravitics Systems* the public was not informed that T.T. Brown had any AC component to his voltage supplies. This disclosure hints at the improvements which the aerospace industry may have been experimenting with to create a Mach 3 aircraft.

Along the same vein, Dr. Cravens, who gave Brown a high rating of "practicality", recently reported in his evaluation of T.T. Brown that older high voltage supplies always had some AC component to the regulated signal, which he wonders if it had any effect in the phenomena.[8] This could be the most valuable part to this report, which is underscored in the following section.

The Hutchison Effect

In 1980, George Hathaway, a professional engineer licensed in Canada, formed a small company to develop and promote what is referred to as *the Hutchison Effect*. It is named after its inventor, John Hutchison, who liked to experiment with combinations of "Tesla coils and Van de Graaff generators" at the same time.

Much of the information about the "lift and disruption" effects has been reported at previous conferences[9]. Videotapes of much of the phenomena have been shown as well. Hathaway recently assembled a three-hour videotape which documents the TV interviews, reports, and actual events.

To summarize, the experiments were conducted with *250 KV* of DC power on the Van de Graaff and about the same of AC power on the Tesla coil. The total real power was about 1.5 KW continuously, according to Hathaway. Besides the disruptive effects, which were numerous, the lifting of various heavy objects by the field was most impressive.

In regards to the AC contribution to the field, Hathaway reports that he measured *2 mV (millivolts) per meter* in the active region (besides the DC offset). This appears to be a small AC signal but on top of the high voltage DC signal, it performs amazing feats.

The 19 lb. bronze bushing experiment remains to this day the most astonishing demonstration of *electric field-induced propulsion* which is the same as externally induced electrogravitics. The frame-by-frame pictures actually were measured and graphed by engineer Hathaway, showing a surprising "linearly increasing acceleration".

All forces in nature are proportional to acceleration, such as in Newton's law. However, the combination of "high "voltage DC with an AC ripple induces a "new type of force" on a relatively heavy object, which exceeds electrostatic effects by several orders of magnitude. I would suggest a term such as "hyperforce" to describe a *linearly increasing acceleration*, which differs from physics terms like "jerk" which is not sustained.

Dr. Robert Flower recently informed me that the Lorenz-Dirac equation is a rare example of a third derivative (d^3x/dt^3) effect similar to the Hutchison effect, which also may be related to the electrogravitics phenomena.

Electric Torsion Pendulum

Thanks to the *Ether Technology* text, Dr. Erwin Saxl's work was brought to the public's attention. In a relatively simple experiment, Dr. Saxl found that an electrically conductive, grounded torsion pendulum behaved strangely during a solar eclipse[10]. The period change was so dramatic that it could be timed to the eclipse precisely.

I had the opportunity to meet with Dr. Saxl's colleague, Mildred Allen from Mt. Holyoke College, a few years ago and obtained an unpublished paper of theirs at that time. The original experiments were performed with a fixed temperature varying less than 0.1C so the criticism that temperature variations could have induced the effects is unfounded.

Dr. Allais, who won a prize from the Gravity Research Foundation in 1959, also confirmed similar results using a conductive paraconical pendulum during a solar eclipse.[11] Alternatively, when the torsion pendulum was charged with high voltage, new phenomena began to emerge. An important relationship is the graph of *voltage* on the pendulum *versus its period*, which is found to be exponentially related. This is equivalent to a 5% change in inertial mass though the maximum variation in g is 10^{-5}.

The fact that gravity and electrostatic fields are related is not unusual, since various physical theories predict it. However, the nature of the interaction is just now being quantitatively analyzed instead of simple qualitative relations. Dr. Saxl notes:

> *When working as a post-doctoral student with Einstein, we discussed the possibility that there were interrelations between electricity, inertial mass and gravitation. These experimental results make me wonder whether they may be so interpreted.*[12]

The B-2 Stealth Bomber Connection

Thanks to Dr. Paul LaViolette reporting in his article, "The U.S. Antigravity Squadron"[13], there is *substantial evidence* that the electrogravitics research of the 1950's actually resulted in the B-2 Stealth Bomber auxiliary propulsion system.

Summarizing Dr. LaViolette's ar-

TABLE OF B-2 BOMBER ELECTROGRAVITIC EVIDENCE

The "crown jewel" of the US Air Force, the B-2 Stealth Bomber incorporates advanced electrogravitic principles!

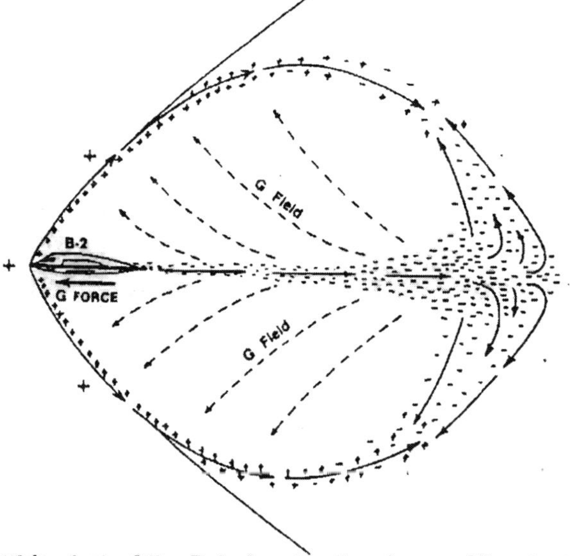

A side view of the B-2 showing the shape of its electrically charged Mach 2 supersonic shock and trailing exhaust stream. Solid arrows show the direction of ion flow; dashed arrows show the direction of the gravity gradient induced around the craft. Credit: Paul LaViolette

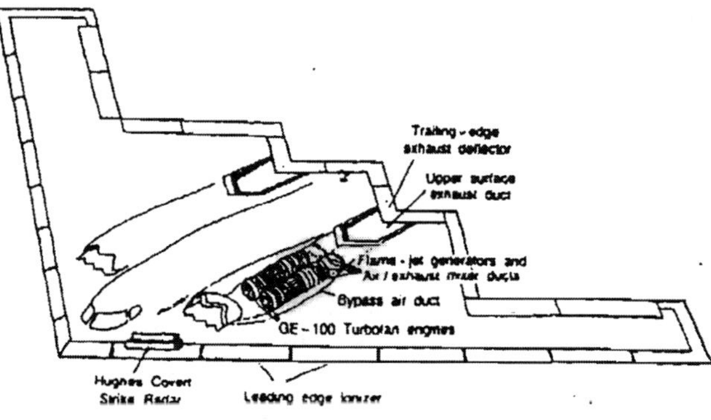

A view of the B-2 with a cutaway showing the arrangement of its flame-jet generators.

The B-2 Stealth Bomber is the "crown jewel" of the United States Air Force. The advanced technology incorporated into the design suggests that the B-2 may be our first step towards a true electrogravitic airship with supersonic capabilities.

Each of the B-2's leading edges is segmented into eight sections separated from one another by ten centimeter wide struts. Quite possibly, the struts electrically isolate the sections from one another so that they may be individually electrified. In this way, through proper control of the applied voltage, it would be possible to gravitically steer the craft. Brown had suggested a similar idea for steering his saucer craft.

The leading-edge sections positioned in front of the air scoops most likely are sparingly electrified so as to prevent positive ions from entering the engine ducts and neutralizing the negative ions being produced there. A summary of the advanced technology and its roots is as follows:

1) The B-2 charges the leading edges of its wing-like body, with high voltage;
2) The B-2 is shaped just like T.T. Brown suggested an electrogravitic craft should look, for maximum charge separation;
3) Northrup tested leading-edge charging in 1968;
4) T.T. Brown suggested that the craft should be powered by a flame-jet generator like the B-2 engine;
5) *Aviation Week* admits the existence of "dramatic, classified technologies" applicable to "aircraft control and propulsion" on the B-2;
6) *Aviation Week* also disclosed that the ceramic RAM on the B-2 outer skin is powdered depleted uranium, which just happens to have a dielectric constant of 3 times that of the high-K dielectrics tested in the 1950's (barium titanate oxide);
7) The B-2's Emergency Power Units (EPU) can work at high altitudes or even in space, driving an electrical generator;
8) Edward Aldridge, the Secretary of the Air Force, admits that the B-2 creates *no vapor trail* at high altitudes.
9) The decomposed gases from the EPU's can function as the ion-carrying medium, according to T.T. Brown.

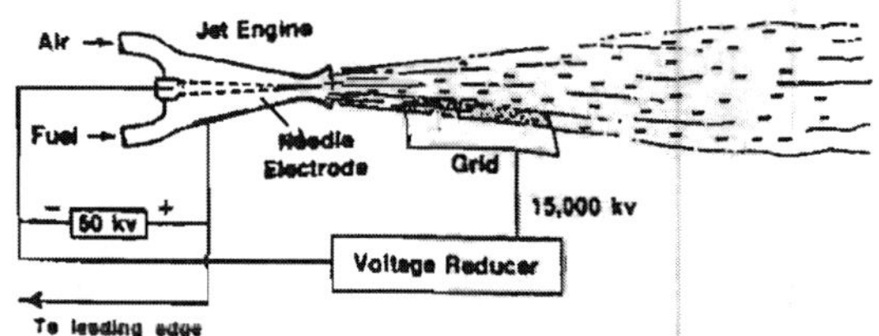

Flame generators were first suggested by T. T. Brown in the 1950s!!! Credit: Paul LaViolette

ticle, with references cited therein, we must emphasize the following table of important highlights. Together, they provide more than enough facts to make the B-2 a more fascinating aircraft than ever realized. Furthermore, the details which Dr. LaViolette uncovered below substantiate his physics theory of *Subquantum Kinetics* (the title of his book) which actually predicts that high voltage will have a propulsion effect when properly polarized.[14]

These exciting facts and more create a sense of excitement about the world's foremost aircraft. Dr. LaViolette argues that the electrogravitic drive will function better at higher speeds due to the shape of the craft causing better gravitic potential. Therefore, it is likely, he says, that the B-2 actually is a supersonic aircraft, especially since the leading-edge charging is primarily for supersonic softening of the shock wave besides electrogravitic capability.

House Representative Robert Walker was quoted in *Popular Science*[15] as promoting the idea of declassifying military secrets that will help commercial development. We hope that this trend will continue so the Spaceplane may also acquire an electrogravitic drive.

It is my belief, based upon T.T. Brown's research and Dr. LaViolette's theories, that an electrogravitic drive is a feasible propulsion method for the Mars journey. It is especially attractive since it uses so little power when it is operational. The complete reports:

- Electrogravitics Systems
- The Gravitics Situation"
- Negative Mass as an Energy Source
- The U.S. Antigravity Squadron

are available in the book, *Electrogravitics Systems: Reports on a New Propulsion Methodology.* The 20th Century has been one of exploring and exploiting the interactions between electricity and magnetism. The 21st Century leads us in a new direction as we learn of the relationship between gravity and electricity. Our electrogravitic future is just beginning! *TV*

References

1) Nipher, *Electrical Experimenter*, March, 1918, p.743
2) Valone, T., ed., *Electro-gravitics Systems: Reports on a New Propulsion Methodology*, Integrity Research Institute, Washington, DC, 1994.
3) Brown, T.T., "How I Control Gravity", *Science and Invention*, 1929
4) *Science and Invention*, "Gravity Nullified", Sept. 1927, p. 398.
5) T.T. Brown U.S. Patents: #1,974,483, #2,949,550, #3,018,394, #3,022,430, #3,187,206 among others.
6) Rho Sigma, *Ether Technology*, 1977.
7) Boylestad, R.L., *Introductory Circuit Analysis*, Merrill Pub., 5th ed., 1987, p. 606. Also see "AC Formula Sheet" pub. by Integrity Res. Inst.
8) Cravens, T.L. "Electric Propulsion Study", AL-TR-89-040, #ADA 227121, Science Applic. Inter. Corp., Torrance, CA 90501
9) "The Hutchison Effect—A Lift and Disruption System", *Proceedings of the Third Inter. Symp. on Non-Conventional Energy Technology*, Hull, Quebec, 1988, pub. as *New Energy Technology*, by PACE, 100 Bronson, Ottawa, Ontario
10) Saxl & Allen, "1970 Solar Eclipse as 'Seen' by a Torsion Pendulum", *Physical Review D*, Vol. 3, No. 4, 1971, p.823. See also, "An Electrically Charged Torque Pendulum", *Nature*, July 11, 1964, p.136 by Dr. Erwin Saxl (lunar eclipse data).
11) Allais, M., "Should the Laws of Gravitation Be Reconsidered?", *Aero/Space Engineering*, Sep 1959, p.46 (Part 2 appears in Oct 1959, p.51)
12) Saxl and Allen, "Observations of Periodic Phenomena with a Massive Torsion Pendulum", unpublished manuscript, 1976
13) LaViolette, Paul, "The U.S. Antigravity Squadron", *Electrogravitics Systems*, Integrity Res. Inst., 1994, p.82, also published in the *Proceedings of 1993 New Energy Symposium*, Denver, Co
14) LaViolette, Paul, *Subquantum Kinetics*, Starburst Foundation; Brown, S., "Secrets of Groom Lake", *Popular Science*, p.54

ELECTRO ~ GRAVITICS

SPACE EXPLORATION

Electrogravitics — an Energy-Efficient Means of Spacecraft Propulsion

(An idea submitted for consideration in NASA's 1990 Space Exploration Outreach Program)

By Paul A. LaViolette, Ph.D., The Starburst Foundation

The proposed propulsion technology would replace the energy-intensive rocket technology presently used for propelling spacecraft. The technology, called "electrogravitics" has already been developed in "black" defense research programs; programs so highly classified that their existence is not publicly acknowledged. Electrogravitics may appear to violate certain assumptions about gravity commonly held by physicists and aeronautical engineers, so the reader is requested to keep an open mind. The technology does exist; it has been under development for the past 40 years; and it has been shown to be feasible both in carefully controlled laboratory experiments and in actual test flights.

Basically electrogravitics is a technology that allows a spacecraft to artificially alter its own gravity field in such a manner that it is able to levitate itself. This is accomplished by applying a megavolt pulsed DC electric potential across the outer hull and wing of the spacecraft. The craft would be designed to have a relatively large body surface area, similar to the flying wing concept employed in the B-2 bomber. Alternatively it could be discoidal in shape with a lenticular cross-section. Thrust would always be in the direction of the craft's positively charged surface. To quote a February 1956 Air Force intelligence report (now declassified), such a craft "can perform the function of a classic lifting surface—it produces a pushing effect on the under surface and a suction effect on the upper, but unlike the airfoil, *it does not require a flow of air to produce the effect.*"[1]

Payoff

The value of this technology is that the craft may achieve Earth orbit flight at a much lower velocity than conventional rocket propulsion and without the huge fuel expenditure. It would eliminate the hazard of polluting the Earth's stratosphere and space environment with aluminum oxide spherules, which has become an increasing problem with the solid fuel boosters currently in use. The fuel requirements for electrogravitic propulsion are *less than one percent* of those presently used to lift the space shuttle into orbit. Problems typically encountered with the Space Shuttle's rocket propulsion technology (e.g., liquid hydrogen leaks, exhaust leaks around O-rings in the solid fuel booster) would not be present in this technology. Due to its much lower power demands, electrogravitics is much safer and more economical.

Performance Characteristics

As early as 1956, an Air Force study estimated that a manned electrogravitic craft could achieve Mach 3 flight capability with a 50,000 kilowatt power requirement. Such airborne electric power generation is within the reach of present technology. It would require two General Electric superconducting generators powered by two 50,000 horsepower rocket turbine engines. The superconducting generators mentioned here were developed for the Air Force in the late 1970's for use in high-altitude aircraft. Incidentally higher efficiencies are achieved in space due to reduced ion leakage from the hull's charged surface.

Other Enabling Technologies

All enabling technologies have been developed. As early as 1958, a small scale model of an electrogravitic powered aircraft was able to lift 110% of its weight. Since then manned vehicles have been secretly developed and are presently being test flown.

Relation to Major Mission Objectives

Electrogravitics would allow NASA to make frequent flights into space without the numerous delays presently plaguing the Space Shuttle launchings. (The present three year wait for repairing the Hubble Telescope could be cut to 3 weeks.) It would allow flights directly from Earth to Mars without the necessity of laboriously constructing a Mars spaceship in Earth orbit. Such a flight would no longer be contingent on the preexistence of a space station. Moreover the high speeds potentially achievable with electrogravitics would allow travel to Mars to be made in under a month.

Previous History

The electrogravitic phenomenon was first discovered by Thomas Townsend Brown in the mid 1920's. In 1929 he published an article that described his preliminary findings about this effect, which later came to be known as the Biefeld-Brown effect.[2] In 1952 Brown demonstrated the electrogravitic propulsion by energizing two 1-1/2 foot diameter disks with 50,000 volts and flying them around a 20 foot diameter course at speeds of 20 miles per hour. A few years later he flew a set of 3 foot diameter disks about a 50 foot diam-

Space Exploration

eter course under a charge of 150,000 volts. The results were so impressive that they were immediately classified.[3]

A considerable amount of historical information on the early development of electrogravitics may be found in the 1956 Air Force intelligence study entitled *Electrogravitics Systems: An examination of electrostatic motion dynamic counterbary and barycentric control.*[4] It defines the term "dynamic counterbary" as "the manipulation of gravitational force lines," and "barycentric control" as "the adjustment of such manipulative capability to produce a stable type of motion suitable for transportation."

In 1952 the Pentagon was presented with a proposal that called for an extensive effort to develop a "Mach 3 combat type disc" by means of a Manhattan District type of project. By September of 1954 the Pentagon had launched a secret government program to develop a manned antigravity craft based on electrogravitic technology. Commenting on this, an October 1954 Air Force intelligence report states:[5]

"...*the indications are now that the Pentagon is ready to sponsor a range of devices to help further knowledge ... Tentative targets now being set anticipate that the first disc should be complete before 1960 and it would take the whole of the 'sixties' to develop it properly, even though some combat things might be available ten years from now... The frame incidentally is indivisible from the "engine." If there is to be any division of responsibility it would be that the engine industry might become responsible for providing the electrostatic energy (by, it is thought, a kind of flame) and the frame maker for the condenser assembly which is the core of the main structure."* (emphasis added)

By 1956 several major aircraft companies had become involved in electrogravitics research. The list includes: Glen Martin, Convair, Sperry-Rand, Sikorsky, Bell, Lear, Clark Electronics, Douglas, Hiller, and General Electric. The electrogravitics report states:[6]

".. in the trade much progress has been made and now most major companies in the United States are interested in counterbary. Groups are being organized to study electrostatic and electromagnetic phenomena. Most of the industry's leaders have made some reference to it. Douglas has now stated that it has counterbary on its work agenda but does not expect results yet awhile. Hiller has referred to new forms of flying platform, Glenn Martin say gravity control could be achieved in six years, but they add that it would entail a Manhattan District type of effort to bring it about. Sikorsky, one of the pioneers, more or less agrees with the Douglas verdict and says that gravity is tangible and formidable, but there must be a physical carrier for this immense trans-spatial force. This implies that where a physical manifestation exists, a physical device can be developed for creating a similar force moving in the opposite direction to cancel it. Clarke Electronics state they have a rig, and add that in their view the source of gravity's force will be understood sooner than some people think General Electric is working on the use of electronic rigs designed to make adjustments to gravity — this line of attack has the advantage of using rigs already in existence for other defense work. Bell also has an experimental rig intended, as the company puts it, to cancel out gravity, and Lawrence Bell has said he is convinced that practical hardware will emerge from current programs. Grover Leoning is certain that what he referred to as an electromagnetic contra-gravity mechanism will be developed for practical use. Convair is extensively committed to the work with several rigs. Lear Inc., autopilot and electronic engineers have a division of the company working on gravity research and so also has the Sperry division of Sperry-Rand. This list embraces most of the U.S. aircraft industry. The remainder, Curtis-Wright, Lockheed, Boeing and North American have not yet declared themselves, but all these four are known to be in various stages of study with and without rigs.

A December 1957 issue of ***Product Engineering*** magazine reported that the Air Force was encouraging research in electrogravitics.[7] A February 1958 issue of ***Business Week*** magazine mentioned the names of a number of companies and institutions backing gravity research, a list that included the Martin Co., Grumman Aircraft, Lockheed, and Sperry-Rand.[8] In January 1955, G.S. Trimble vice-president of advanced design for Glenn Martin was quoted as saying:[9]

"Unlimited power, freedom from gravitational attraction, and infinitely short travel time are now becoming feasible."

Also Dr. Walter Dornberger, a guided missile consultant for Bell Aircraft, predicted that airliners would eventually travel at speeds of 10,000 miles an hour (Mach 13).

By 1958 Townsend Brown had succeeded in developing a saucer model that, when energized with between 50,000 to 250,000 volts of direct current, was capable of lifting itself up with a thrust equalling *110 percent* of its weight.[10,11] Since that time manned vehicles using the electrogravitics principle have been built and flown.

How to Proceed in Discovering Information on Electrogravitics

It would be worth contacting some of the companies mentioned above. Whether they disclose much about what is going on is another question. Seeing the benefits to NASA of using this technology, perhaps the federal government will realize that the time is ripe

Space Exploration

to begin declassifying this technology. All along, the aircraft industry has been interested in commercializing the technology. So the opportunity of developing a new propulsion system for NASA might serve as the needed catalyst to bring this technology out of the closet and into public use.

If this approach initially fails, NASA officials might try exerting political pressure on the appropriate government agencies to bring about a change in the technology's classification. The National Security Agency might be a good place to start. It is uncertain whether the Pentagon or Whitehouse would have much say in this matter. There is no point in contacting the director of the Air Force Office of Scientific Research, since he is not likely to have been briefed on the existence of this "black" technology. NASA might begin by putting together a group to gather information about electrogravitics and to lobby for a change in its classification status. There are people presently working at NASA who would be good candidates for such a group and who would be knowledgeable about whom to contact. I could provide the name of at least one person who could be helpful.

The Technology's Likelihood of Success and the Classification Issue

Until more is known, it would be premature to say whether there are any outstanding engineering problems in applying the technology to space travel. The main problem appears to be political, rather than engineering in nature. Namely, the need to change the government's present stance on keeping the technology under wraps. With the present move toward the democratization of the Soviet Union, cold war arguments that formerly were used to justify keeping this technology secret, now more than ever, would be treading on thin ice. There is some speculation that the B-2 stealth bomber utilizes electrogravitics technology. If so, perhaps now that a fleet has been built and test flown, the military sector may feel that it is one step ahead of everybody and be more willing to release some information.

To allay fears about keeping antigravity technology classified, one could point out that the automobile probably posed a similar technological advance at the turn of the century. Although it could be argued that gasoline powered vehicles had an advantage over the horse and buggy in that they allowed enemy troops to mobilize a more forceful attack. On the other hand, it might also be argued that the same technology made it possible for defensive forces to effectively ward off attacks. So in the longrun the automobile was not politically destabilizing, although it did bring about major changes in the structure of society. The same could be said to be true of electrogravitics.

Estimated Cost

Because of the difficulty of finding out cost information on this secret technology, it is not possible at this time to make a definite estimate of the cost to construct an electrogravitic spacecraft. Perhaps $500 to $600 million might be in the right ball park. Considering that the vehicle does not require the use of expendable booster rockets, a considerable savings in operating cost would be realized after just a few flights.

Applications Beyond Space Exploration

Electrogravitics holds great benefits for society, beyond its obvious ability to revolutionize space exploration. Taking an historical perspective, it may be seen that mankind's standard of living has accelerated following the introduction of innovations in three principle areas: energy production, transportation, and communications. In the terrestrial transportation area, the widespread use of gravity control technologies would revolutionize our way of life. Commuters would be able to travel vertically as well as horizontally. Rush hour traffic jams would be a thing of the two-dimensional past. Although, increased air traffic would necessitate the introduction of technologies to prevent midair collisions. T·ansport speeds would be vastly increased with considerable economy in fuel consumption. Antigravity vehicles would revolutionize farming, mining, building construction, and shipping, stimulating the world economy beyond our wildest dreams.

Electrogravitics would bring the world closer together. It would erase the distance barriers that currently separate the nations of the world. The ability to travel from New York to Perth, Australia in one hour would make the peoples of the world more internationally oriented. With the ability of cheap high speed travel, national boundaries would begin to dissolve. People would begin addressing world problems from a more planetary perspective, rather than from a regional perspective, thereby aiding the development of world peace.

Electrogravitics would also bring about a new mode of energy production. By arranging electrogravitic capacitor elements around the spokes of a wheel, all aimed in the same rotary direction, when energized they would create a rotational gravity field, which in turn would cause the wheel to turn like a pin wheel. In effect, the wheel would rotate in a state of circular free fall. A device of this sort was patented by Brown in 1934 (U.S. #1,974,483) and tested in 1955 in vacuum chamber experiments he conducted in Paris under the auspices of a French aeronautics corporation. When energized with 200,000 volts, the speed of his rotor began to increase unchecked, reaching such a high rotational velocity that the voltage had to be reduced to keep the rotor from flying apart. An ion wind could not be responsible for this propulsion since the air pressure in the vacuum chamber was less than one billionth of an atmosphere. By con-

Space Exploration

necting the rotor's shaft to a generator, theoretically the rotor could generate more electric power than it consumed, thereby supplying an unlimited source of energy. Hazards such as are typically associated with fossil fuel and nuclear energy generation technologies would be avoided. Given the impending global CO_2 build-up an argument could be made that it is a matter of human survival that we declassify and commercialize this technology as soon as possible.

Related Gravity Technologies

Several approaches to gravity control have been proposed. Some background may be learned by perusing the patent literature. Townsend Brown had applied for numerous patents on his electrogravitic propulsion designs. His U.S. patent #3,187,206 filed in 1958 and granted in 1965 is of particular interest. However, the patents do not give any details on the variable frequency pulsed DC power supply which is a key to the technology. Also of interest are patents by Agnew Bahnson: #2,958,790 (1960), #3,223,038 (1965), #3,227,901 and #3,263,102 (1966). After they were issued, the rights to them were a hot item for purchase by certain corporations who sold and resold them to companies whose names are kept confidential. Henry Wallace presents a different approach to gravity control which utilizes mechanical motion. This is disclosed in his U.S. patents: #3,626,605 and #3,626,606 (1971). Sandy Kidd of Scotland also has devised a mechanical gyroscopic device that is able to achieve a state of complete levitation. He is now developing this technology in Dandenong, Victoria in northwestern Australia, under the sponsorship of millionaire Noel Carrol.

In summary, if NASA were to push for the development of gravity control technologies, they would not only be moving closer to a means of efficiently accomplishing their own objectives, but they would also be doing a great service to humanity.

References

1. ***Electrogravitics Systems: An examination of electrostatic motion, dynamic counterbary and barycentric control.*** *Report No. 13. Aviation Studies (International) Ltd., Special Weapons Study Unit, London, February 1956, pp. 34. (Library of Congress No. 3,1401,00034,5879; Call No. TL565.A9).**
2. *Brown, T. T. "How I control gravity."* ***Science and Invention Magazine****. August 1929.*
3. *Burridge, G. "Another step toward anti-gravity."* ***The American Mercury*** *86(6) (1958):77-82.*
4. ***Electrogravitics Systems****. op. cit., p. 19.*
5. *Ibid., pp. 25 - 27.*
6. *Ibid., pp. 9 - 10.*
7. *"Electrogravitics: Science or daydream?"* ***Product Engineering****. December 30, 1957, p. 12.*
8. *"How to 'fall' into space."* ***Business Week****. Feb. 8, 1958, pp. 51- 53.*
9. *Keyhoe, Major D. E.* ***The Flying Saucer Conspiracy****. New York: Henry Holt & Co., 1955, pp. 251-252.*
10. *"The gravitics situation." Gravity Rand Ltd., December 1976, p.4.*
11. *Brown, T. T., Letter dated February 9,1982.*

* *Only one library has an original copy of this. This is kept at the Wright-Patterson Air Force Base Technical Library in Dayton, Ohio.*

ELECTRO ~ GRAVITICS

Empirical Analysis of Electrogravitics and Electrokinetics and its Potential for Space Travel

Thomas F. Valone*
Integrity Research Institute, Beltsville MD 20705
www.IntegrityResearchInstitute.org
e-mail: IRI@starpower.net

An analysis of the 90-year old science of electrogravitics (a.k.a. "gravitics" or "electrogravity") necessarily includes an analysis of electrokinetics. Electrogravitics is most commonly associated with the 1928 British patent #300,311 of Thomas Townsend Brown (his first patent), the 1952 Special Inquiry File #24-185 of the Office of Naval Research into the "Electro-Gravity Device of Townsend Brown" and two widely circulated 1956 Aviation Studies Ltd. reports on "Electrogravitics Systems", "The Gravitics Situation" and original experiments by Professor N. Nipher in 1918. By definition, electrogravitics historically has had a purported relationship to gravity or the object's mass, as well as the applied voltage. The Gravitics Situation report defined electrogravitics as "The application of modulating influences on electrostatic propulsion system." It also was tested recently by T. Musha of the Honda Corporation, which published experimental results and proposed theory of a correlation between electricity and gravity. Electrokinetics, on the other hand, is more commonly associated with many later patents of T. Townsend Brown as well as Agnew Bahnson, starting with the 1960 US patent #2,949,550 entitled, "Electrokinetic Apparatus." Electrokinetics, which often involves a capacitor and dielectric, has virtually no relationship that can be connected with mass or gravity. The Army Research Lab has recently issued a report on electrokinetics, analyzing the force on an asymmetric capacitor, while NASA has received three patents on the same design topic. To successfully describe and predict the reported motion toward the positive terminal of the capacitor, it is desirable to use the classical electrokinetic field and force equations for the specific geometry involved. This initial review and analysis also suggests directions for further confirming experiments and an empirically-based formulation of a working hypothesis for electrokinetics.

I. Nomenclature

J	=	electric current density
I	=	electric current
E_K	=	electrokinetic force vector
$\mathbf{B}$	=	magnetic flux density
$\mathbf{E}$	=	electric field
ρ	=	charge density

II. Introduction to Electrogravitics and Electrokinetics

FOURTEEN years ago the first edited volume on the subject, *Electrogravitics Systems Volume I: A New Propulsion Methodology* or just "Volume I", introduced the subject by reprinting the Aviation Studies reports from 1956 as well as an in-depth analysis of the B-2 bomber by Paul LaViolette.[1] The second volume, *Electrogravitics II: Validating Reports on a New Propulsion Methodology* or "Volume II" expands the historical perspective of the first volume and updates it (ref. 3). For example, Volume II contains further information on the Army Research Lab and Honda Corporation experiments, as well as the electrokinetic equation discovery presented in this paper. A short review of the history of electrogravitics has recently been published by Professor Theodore Loder.[2]

---

* President, Integrity Research Institute, 5020 Sunnyside Avenue, Suite 209, Beltsville MD 20705, AIAA Member.

A working definition, based on the T. Townsend Brown's first patent #300,311 and The Gravitics Situation report is "*electricity used to create a force that depends upon an object's mass, similar to gravity.*" This is the answer that perhaps should still be used to identify true electrogravitics, which also involves the object's mass in the force, often with a dielectric. This is also what the "Biefeld-Brown effect" of describes. However, we have seen T. Townsend Brown and his patents evolve over time which Tom Bahder emphasizes. Later on, Brown refers to "electrokinetics" (that partly overlaps the field of electrogravitics), that requires asymmetric capacitors to amplify the force. Therefore, Bahder's article discusses the lightweight effects of "lifters" and the ion mobility theory found to explain them. Note: *electrogravitics (EG) and electrokinetics (EK) are related but different phenomona.*

To put things in perspective, the article "How I Control Gravitation," published in 1929 by Brown,[3] presents an electrogravitics-validating discovery about *very heavy metal objects* (44 lbs. each) separated by an insulator, charged up to high voltages. T.T. Brown also expresses an experimental formula in words which tell us what he found was directly contributing to the *unidirectional force* (UDF) which he discovered, moving the system of masses toward the positive charge. He describes the equation for his electrogravitic force to be $F \approx Vm_1m_2/r^2$. However, electrokinetics and electrogravitics also seem to be governed by another equation (Eq.1) when higher order pulsed voltages are utilized .

A. Zinsser Effect versus the Biefeld-Brown Effect

To expand and support the empirical evidence for electrokinetics, there is another invention which has comparable experiments that also involve electrogravity, called "gravitational anisotropy" by Rudolf G. Zinsser from Germany. Zinsser presented his experimental results at the Gravity Field Conference in Hanover in 1980, and also at the First International Symposium of Non-Conventional Energy Technology in Toronto in 1981.[4] For years afterwards, all of the scientists who knew of Zinsser's work, including myself, regarded his invention as a unique phenomenon, not able to be classified with any other discovery. However, upon comparing Zinsser to Brown's 1929 article on gravitation referred to above, there are striking similarities.

Zinsser's discovery is detailed in *The Zinsser Effect* book by this author.[5] To summarize his life's work, Zinsser discovered that if he connected his patented pulse generator to two conductive metal plates immersed in water, he could induce a sustained force that lasted even after the pulse generator was turned off. The pulses lasted for only a few nanoseconds each.[6] Zinsser called this input "a kinetobaric driving impulse." Furthermore, he points out in the Specifications and Enumerations section, that the high dielectric constant of water (about 80) is desirable and that a solid dielectric is possible. Dr. Peschka calculated that Zinsser's invention produced 6 Ns/Ws or 6 N/W.[7] This figure is *twenty times* the force per energy input of the Inertial Impulse Engine of Roy Thornson, (report available from IRI) which has been estimated to produce 0.32 N/W.[8] By comparison, it is important to realize that any production of force today is less efficient, as seen by the fact that a DC-9 jet engine produces *about 20 times less:* only 0.016 N/W or 3 lb/hp (fossil-fuel-powered land and air vehicles are even worse.)

Let's now compare the Zinsser Effect with the Biefeld-Brown Effect, looking at the details. Brown reports in his 1929 article that there are effects on plants and animals, as well as effects from the sun, moon and even slightly from some of the planetary positions. Zinsser also reports beneficial effects on plants and humans, including what he called "bacteriostasis and cytostasis."[9] Brown also refers to the "endogravitic" and "exogravitic" times that were representative of the charging and discharging times. Once the gravitator was charged, depending upon "its gravitic capacity" any further electrical input had no effect. *This is the same phenomenon that Zinsser witnessed* and both agree that the *pulsed voltage generation* was the main part of the electrogravitic effect.[10] Both Zinsser and Brown worked with dielectrics and capacitor plate transducers to produce the electrogravitic force. Both refer to a high dielectric constant material in between their capacitor plates as the preferred type to best insulate the charge. However, Zinsser never experimented with different dielectrics nor higher voltage to increase his force production. This was always a source of frustration for him but he wanted to keep working with water as his dielectric.

B. Electrically Charged Torque Pendulum of Erwin Saxl

Brown particularly worked with a torque (torsion) pendulum arrangement to measure the force production. He also refers the planetary effects being most pronounced *when aligned with the gravitator* instead of perpendicular to it. He compares these results to Saxl and Allen, who worked with an electrically charged torque pendulum.[11] Dr. Erwin Saxl used high voltage in the range of +/- 5000 volts on his very massive torque pendulum.[12] The changes in period of oscillation measurements with solar or lunar eclipses, showed great sensitivity to the shielding effects of gravity during an alignment of astronomical bodies, helping to corroborate Brown's observation in his 1929 article. The pendulum Saxl used was over 100 kilograms in mass.[13] Most interesting were the "unexpected phenomena" which Saxl reported in his 1964 *Nature* article (see ref. 11). The positively charged pendulum had the longest period of oscillation compared to the negatively charged or grounded pendulum. Dirunal and seasonal variations were found in the effect of voltage on the pendulum, with the most pronounced occurring during a solar or lunar eclipse. In my opinion, this demonstrates the basic principles of electrogravitics: high voltage and mass together will cause unbalanced forces to occur. In this case, the electrogravitic interaction was measurable by oscillating the mass of a charged torque pendulum (producing current) whose period is normally proportional to its mass.

C. Electrogravitic Woodward-Nordtvedt Effect

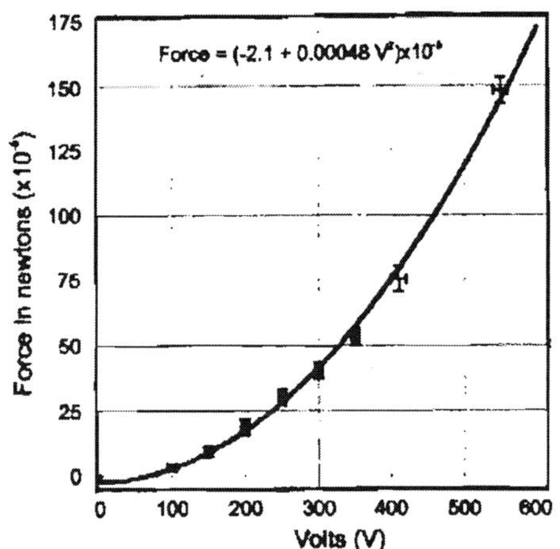

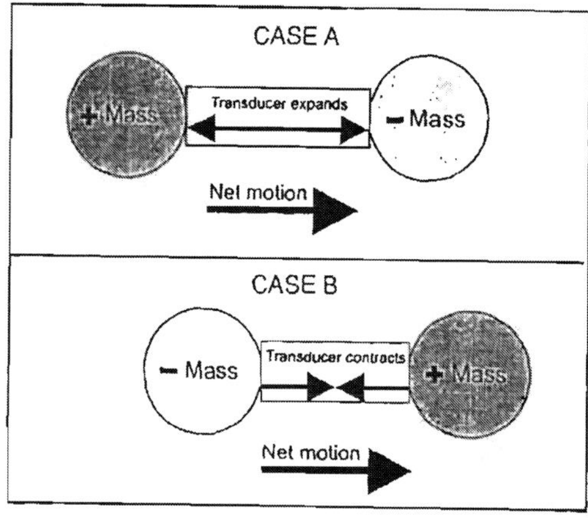

Figure 1. Force Output Vs. Capacitor Voltage Input of a Woodward Force Transducer (Mahood, 2000) and the Net Motion Direction of Cases A and B (Woodward, 2000). *Reported data graph of the Woodward-Nordtvedt effect. Note that the reported force is Newtons ($x10^{-5}$) which equals dynes)*

Referring to mass, it is sometimes not clear whether gravitational mass or inertial mass is being affected. The possibility of altering the equivalence principle (which equates the two), has been pursued diligently by Dr. James Woodward[14] (patent cover sheets in Volume II). His prediction, based on Sciama's formulation of Mach's Principle in the framework of general relativity, is that "in the presence of *energy flow*, the inertial mass of an object may undergo sizable variations, changing as the 2nd time derivative of the energy."[15] Woodward, however, indicates that it is the "active gravitational mass" which is being affected but the equivalence principle causes both "passive" inertial and gravitational masses to fluctuate.[16] With barium titanate dielectric between disk capacitors. a 3 kV signal was applied in the experiments of Woodward and Cramer resulting in symmetrical mass fluctuations on the order of centigrams.[17] Cramer actually uses the phrase "Woodward effect" in his AIAA paper, though it is well-known that Nordtvedt was the first to predict noticeable mass shifts in accelerated objects.[18]

The interesting observation which can be made, in light of previous sections, is that Woodward's experimental apparatus *resembles a combination of Saxl's torsion pendulum and Brown's electrogravitic dielectric capacitors.* The differences arise in the precise timing of the pulsed power generation and with input voltage. Recently, 0.01 μF capacitors (Model KD 1653) are being used, in the 50 kHz range (lower than Zinsser's 100 kHz) with the voltage still below 3 kV. Significantly, the thrust or unidirectional force (UDF) is exponential, depending on the square of the applied voltage.[19] However, the micronewton level of force that is produced *is actually the same order of magnitude which Zinsser produced,* who reported his results in dynes (1 dyne = 10^{-5} Newtons).[20] Zinsser had *activators* with masses between 200 g and 500 g and force production of "100 dynes to over one pound."[21] Recently, Woodward has been referring to his transducers as "flux capacitors" (like the movie, *Back to the Future*).[22]

III. Jefimenko's Electrokinetics Equations

Known for his extensive work with atmospheric electricity, electrostatic motors and electrets, Dr. Oleg Jefimenko deserves significant credit for presenting a valuable theory of the *electrokinetic field*, as he calls it.[23] A W.V. University professor and physics purist at heart, he describes this field as the *dragging force* that electrons exert *on neighboring electric charges*, which is what he says Faraday noted in 1831, when experimenting with parallel wires: a momentary current in the same direction when the current is turned on and then a reverse current in the adjacent wire when the current is turned off.

He identifies the *electrokinetic field* by the vector $\mathbf{E}_k$ where

$$\mathbf{E}_k = -\frac{1}{4\pi\varepsilon_o c^2}\int \frac{1}{r}\left[\frac{\partial \mathbf{J}}{\partial t}\right]dv' \qquad (1)$$

It is one of three terms for the electric field in terms of current and charge density. Equations like $\mathbf{F} = q\mathbf{E}$ also apply for calculating force. The significance of $\mathbf{E}_k$, as seen in Eq. 1, is that the electrokinetic field simply the third term of a classical solution for *the electric field* in Maxwell's equations:

$$\mathbf{E} = \frac{1}{4\pi\varepsilon_o}\int \left\{\frac{\rho}{r^2} + \frac{1}{rc}\frac{\partial \rho}{\partial t}\right\}\mathbf{r}\,dv' + \mathbf{E}_k \qquad (2)$$

This three-term equation is a causal equation, according to Jefimenko, because it links the electric field $\mathbf{E}$ back the electric charge and its motion (current) which induces it. (He also proves that $\mathbf{E}$ cannot be a causal consequence of a time-variable magnetic field $\partial\mathbf{B}/\partial t$ but instead occurs simultaneously.) This is the essence of electromagnetic induction, *as Maxwell intended*, which is measured by, not caused by, a changing magnetic field. The third electric field term, designated as the electrokinetic field, *is directed along the current direction or parallel to it.* It also exists only as long as the current is changing in time. Lenz' Law is also built into the minus sign. Parallel conductors will produce the strongest induced current.

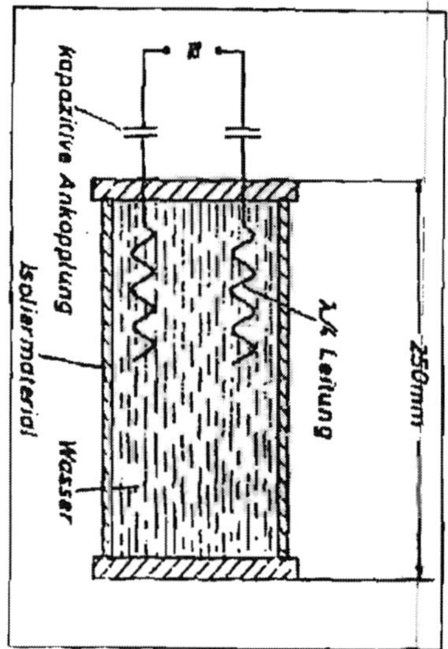

Figure 2. Sample capacitor probe used by Zinsser. *Note the quarter λ/4 wavelength electrodes that indicate an electrically resonant circuit design.*

The significance of Eq. 3 is that the magnetic vector potential is seen to be created by the time integral which amounts to an *electrokinetic impulse* "produced by this current at that point *when the current is switched on*" according to Jefimenko.[24] Of course, a time-varying sinusoidal current will also qualify for production of an electrokinetic field and the vector potential. An important consequence of Eq. 1 is that *the faster the rates of change of current, the larger will be the electrokinetic force*. Therefore, high voltage pulsed inputs are favored.

However, its significance is much more general. "This field can exist anywhere in space and can *manifest itself as a pure force* by its action on free electric charges." All that is required for a measurable force *from a single conductor* is that the change in current density (time derivative) happens very fast (the c^2 in the denominator is also equal to $1/\mu_o\varepsilon_o$ unless the medium has non-vacuum permeability or permittivity).

The electrogravitics experiments of Brown and Zinsser involve a dielectric medium for greater efficacy and charge density. The electrokinetic force on the electric charges (electrons) of the dielectric, according to Eq. (1), is in the *opposite direction of the increasing positive current* (taking into account the minus sign). For parallel plate capacitors, Jefimenko explains that *the strongest induced field is produced between the plates* and so another equation evolves.

IV. Electrokinetic Force Predicts Propulsion Direction

Can Jefimenko's electrokinetic force empirically and qualitatively predict the correct *direction* of the electrogravitic force seen in the Zinsser, Brown, Woodward as well as the yet-to-be-discussed Campbell, Serrano, and Norton AFB craft demonstrations? The following four sections offer empirical evidence for a "prediction" of a force production direction.

1) Starting with *Zinsser's probe diagram* (Fig. 2) from Prof. Peschka's article, it is purposely put on its end in order to compare it with an equivalent parallel plate capacitor (the plates are x distance apart) from Jefimenko's book:[25] Professor Jefimenko performs a calculation of the electrokinetic force in the space between two current-carrying capacitor plates powered by an alternating current. He designates X for the space between the plates where W is the width of each plate and the height is not labeled. His example matches the Zinsser force transducer quite closely.

We note that the current is presumed to be the same in each plate but in opposite directions because it is alternating. Using $E = -\partial A/\partial t$, Jefimenko calculates the electrokinetic field, for the AC parallel plate capacitor with current going in opposite directions, as

$$E_k = -\mu_o \frac{\partial I}{\partial t} \frac{x}{w} \mathbf{j} \qquad (3)$$

where **j** is the unit vector for the y-axis direction. It is also assumed that the y-axis points upward in Fig. 2 and so with the minus sign of Eq. 3, the electrokinetic force for the AC parallel plate capacitor *will point downward.* Since Zinsser had his torsion balance on display in Toronto in 1981, I was privileged to verify the direction of the force that is created with his quarter-wave plates oriented as they are in Fig. 2. His torsion balance is built so that the capacitor probe can only be deflected *downward* from the horizontal. *The electrokinetic force is in the same direction.*

2) Looking at *Brown's electrogravitic force direction* from Fig. 3 in his 1929 article "How I Control Gravitation," we see that the positive lead is on the right side of the picture. Also, the arrow below *points to the right* with the caption, "Direction of movement of entire system toward positive." Examining the electrokinetic force of Eq. 1 in this article, we note that the increasing positive current comes in by convention in the positive lead and points to the left. Therefore, considering the minus sign, the direction of the electrokinetic force will be *to the right*. Using the same analysis on Brown's 1929 article results in *confirmation of induced electrokinetic force direction.*[26] Thus, with Zinsser's and Brown's gravitators, *the electrokinetic theory provides a useful explanation and it is accurate for prediction of the resulting force direction.*

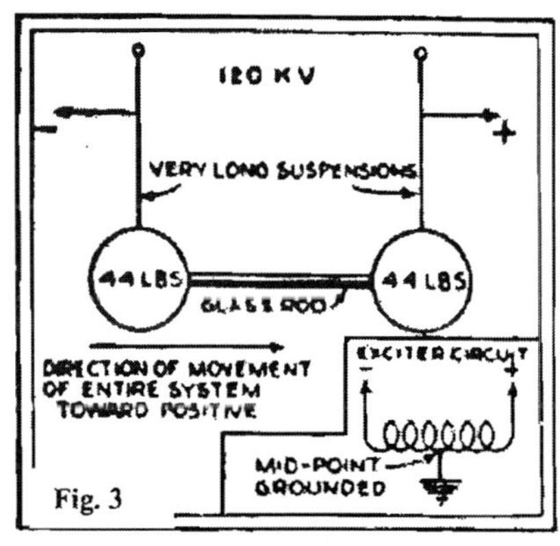

A SIMPLE TYPE OF GRAVITATOR IS SHOWN IN THE ABOVE ILLUSTRATION.

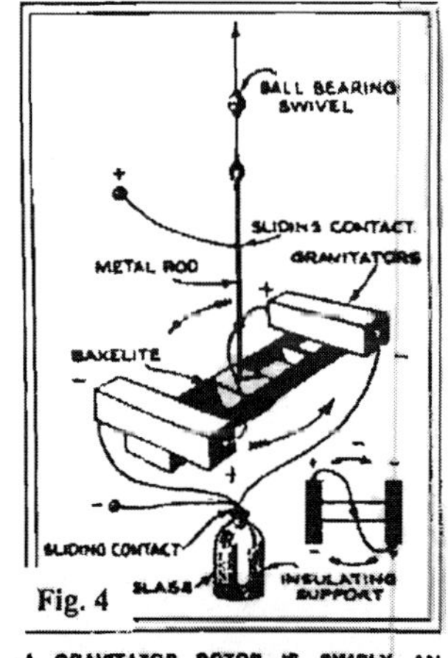

A GRAVITATOR ROTOR IS SIMPLY AN ASSEMBLY OF UNITS SO MADE THAT ROTATION RESULTS UNTIL THE IMPULSE IS EXHAUSTED.

It is also worthwhile noting that T.T. Brown also indicates in that article,

> "when the direct current with high voltage (75 – 300 kilovolts) is applied, the gravitator swings up the arc ... but it does not remain there. The pendulum then gradually returns to the vertical or starting position, even while the potential is maintained...Less than five seconds is required for the test pendulum to reach the maximum amplitude of the swing, but from thirty to eighty seconds are required for it to return to zero."

This phenomenon is *remarkably the same type of response that Zinsser recorded* with his experimental probes. Jefimenko's theory helps explain the rapid response, since the change of current happens in the beginning. However, the slow discharge in both experiments (which Zinsser called a "storage effect") needs more consideration. Considering the electrokinetic force of Eq. 3 and the +/- derivative, we know that the slow draining of a charged capacitor, most clearly seen in Fig. 3 of Brown's 1929 article, will produce a decreasing current out of the + terminal (to the right) and in Eq. 3, this means the derivative is negative. Therefore, *the slow draining of current will produce a weakening electrokinetic force* but *in the same direction as before*! The force will thus sustain itself to the right during discharge, explains the slow return.

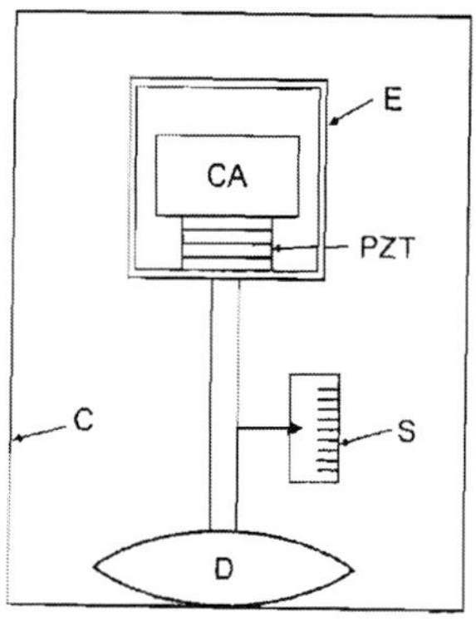

Figure 5. Woodward's #6,098,924 patented impulse engine, also called a "flux capacitor." *The PZT provides nanometer-sized movements that are timed to an AC signal input. A torsion balance has been used*

3) It is reasonable at this stage to also suggest that the electrokinetic theory will also predict the direction of *Woodward's UDF* but instantaneous analysis needs to be made to compare current direction into the commercial disk capacitors and the electrokinetic force on the dielectric charges. In every electrogravitics or electrokinetics case, it can be argued, the "neighboring charges" to a capacitor plate will necessarily be those in the dielectric material, which are polarized. The bound electron-lattice interaction *will drag the lattice material with them*, under the influence of the electrokinetic force. If the combination of physical electron acceleration (which also can be regarded as current flow) and the AC signal current flow can be resolved, it may be concluded that an instantaneous electrokinetic force, depending on dI/dt, contributes to the Woodward-Nordtvedt effect.

4) The *Campbell and Serrano capacitor modules* seen in their patented drawings in Figs. 6 and 7, as well as the Fig. 9 hovercraft *demonstration unit (Norton AFB, 1988)*,[27] can also be analyzed with the electrokinetic force, in the same way that the Brown gravitator force was explained in paragraph (2) above. The current flows in one direction through the capacitor-dielectric and the force is produced in the opposite direction. The Norton AFB electrokinetic craft of Fig. 9 has big 12 foot plate sections but the current flow still starts at the center, *through the plates*. The Serrano patent diagram is also very similar in construction and operation. Campbell's NASA patents include #6,317,310, #6,411,493, and #6,775,123.

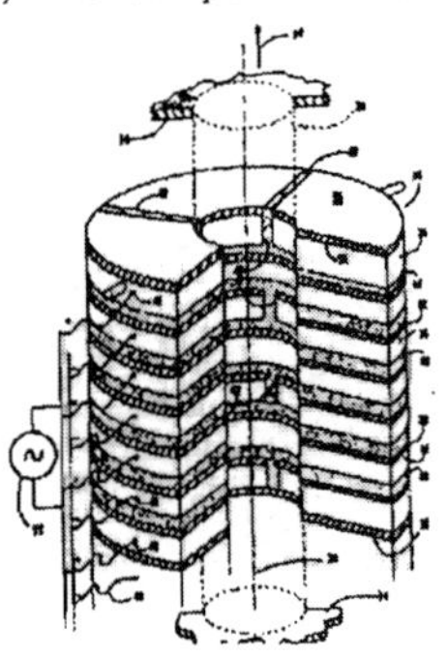

Figure 7. Capacitor propulsion device. *alternating metal and dielectric layers from Serrano's PCT patent WO 00/58623 with upward thrust direction indicated and + and – polarity designated on the side.*

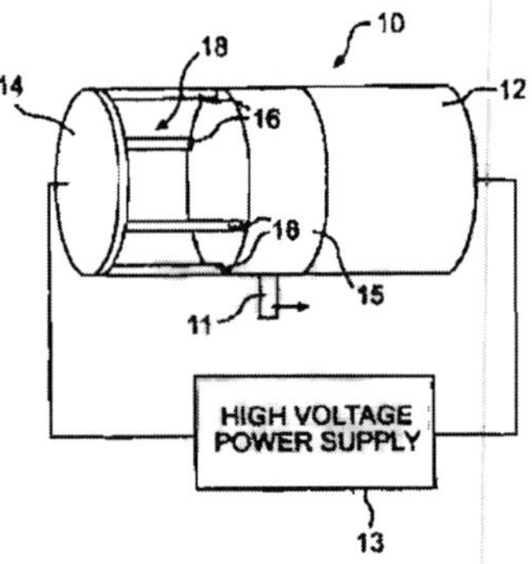

Figure 6. Capacitor module from Campbell's NASA patent #6,317,310 which creates a thrust force. Disk 14 is copper; Struts 16 are dielectrics; Cylinder 15 is a dielectric; Cylinder 12 is an axial capacitor plate; Support post 11 is also dielectric.

V. Electrokinetic Theory Observations

For parallel plate capacitor impulse probes, like Zinsser, Serrano, Campbell, the Norton AFB craft and both of Brown's models, the electrokinetic field of Eq. 3 provides a working model that seems to predict the *nature and direction of the force during charging and discharging phases.* More detailed information is needed for each example in order to actually calculate the theoretical electrokinetic force and compare it with experiment. We note that Eq. 3 also does not suffer the handicap of Eq. 1 since no c^2 term occurs in the denominator. Therefore, it can be concluded that AC fields operating on parallel plate capacitors should create *significantly larger* electrogravitic forces than other geometries with the same dI/dt. However, the current I is usually designated as $I_o\sin(\omega t)$ and its derivative is a sinusoid as well. Therefore, a detailed analysis is needed for each specific circuit and signal to determine the outcome.

Eq. 3 also seems to suggest a *possible enhancement* of the force *if a permeable dielectric (magnetizable) is used.* Then, the value for μ of the material would normally be substituted for μ_o.[28]

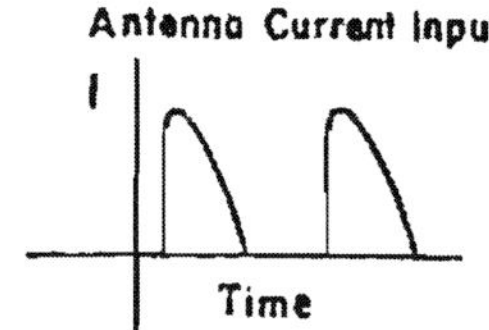

Figure 8. A possible electrokinetic force current waveform. *Schlicher propulsion patent #5,142,861*

A further observation of both Eq. 1 and Eq. 3 is that very fast changes in current, such as *a current surge or spark discharge* has to produce the most dynamic electrokinetic force, since dI/dt will be very large.[29] *The declining current surge,* as in Section 2, will simply decrease the force, with an opposing force when the current reverses direction. *Creative waveshaping seems to be the answer* to this obvious dilemma. Fortunately, a few similar inventions use pulse power electric current generators to create propulsion. The Taylor patent #5,197,279 "Electromagnetic Energy Propulsion Engine" uses huge currents to produce magnetic field repulsion. The Schlicher patent #5,142,861 "Nonlinear Electromagnetic Propulsion System and Method" predicts hundreds of pounds of thrust with tens of kiloamperes input. The Schlicher antenna current input is a rectified current surge produced with an SCR-triggered DC power source (see Fig. 8). The resulting waveform has a very steep leading edge but a *slowly declining trailing edge*, which should also be desirable for the electrokinetic force effect.[30] Furthermore, if this waveform is continued into the negative current direction below the horizontal axis, all of that region reinforces the electrokinetic force, with no opposite forces. Therefore, *a complete sinusoidal wave*, with Schlicher-style steep rise-times is recommended for a signal that contributes to a unidirectional force during 75% of its cycle.

Another observation that should be mentioned is that this electrokinetic force theory does not include the mass contribution to the electrogravitic force which Saxl, Woodward, and Brown's 1929 gravitator emphasize. A contributor to *Electrogravitics II*, Takaaki Musha offers a derived equation for electrogravitics *that does include a mass term* but not a derivative term. His model is based on the charge displacement or "deformation" of the atom under the influence of a capacitor's 18 kV high voltage field and his experimental results are encouraging. He also includes a reference to Ning Li and her *gravitoelectric theory.*[31]

A final concern, which may arise from the very nature of the electrokinetic force description, is the difficulty of conceptualizing or simply accepting the possibility of an *unbalanced force creation pushing against space.* This author has wrestled with this problem in other arenas for years. Three examples include (1) the homopolar generator which creates *back torque* that ironically, *pushes against space* to implement the Lorentz force to slow down the current-generating spinning disk.[32] Secondly (2), there is the intriguing *spatial angular momentum discovery* by Graham and Lahoz.[33] They have shown, reminiscent of Feynman's "disk paradox," that the vacuum is the seat of Newton's third law. A torsion balance is their chosen apparatus as well to demonstrate the pure reaction force with induction fields. Their reference to Einstein and Laub's papers cites the time derivative of the Poynting vector $\mathbf{S} = \mathbf{E} \times \mathbf{H}$ integrated over all space to preserve Newton's third law. Graham and Lahoz predict that *magnetic flywheels with electrets* will circulate energy to *push against space.* Lastly, for (3), the Taylor and Schlicher inventions push against space with an unbalanced force that is electromagnetic in origin.

A further confirmation of an electromagnetic explanation for the electrokinetic force empirically can be found in the semiconductor integrated circuit industry. Bothra's US patent #6,191,481 describes an electromigration impeding metallization lines and oxide slots that purposely cause "back-flow" (col. 6, line 25-30). The back-flow of electrons literally causes a force that not only stops electromigration, but if large enough, may perhaps be argued to cause a transfer of momentum to the lattice. This is a direction for high amperage pulsed current experiments to consider for a theoretical foundation for the propulsive force production.

At the Utah chapter meeting of the National Space Society in 2006, a military contractor also described his work with asymmetric capacitors which were summarized as "levitating a hockey puck" with pulsed currents.

VI. Eye Witness Testimony of Advanced Electrogravitics

Sincere gratitude is given to Mark McCandlish who, though intimitated for publicizing this military artwork, offers us one of the most conclusive rendition of a covert, flat-bottomed saucer hovercraft seen by dozens of invited eye-witnesses, including a Congressman, at Norton AFB on November 12, 1988. Contacting Dr. Hal Puthoff about Mark's story, shortly after the famous Disclosure Event[34] at the National Press Club in 2001, he explained to me that he had already performed due diligence on it and checked on each individual to verify the details of the story. Hal explains,

> "All I was able to determine by my due diligence was: (1) to independently interview the source of the story and verify that, indeed he did tell the story to the individual who had passed it on to me, and (2) to independently interview yet another individual who had heard a similar story from a separate source. BUT, I was never able to verify that the story itself was true, only that there were two individuals who said it was true. I then corrected you with my statement (exact quote): '... the story remains in my 'gray basket' only as 'possibly' true.'"

Since Dr. Puthoff used to work for the CIA for ten years as a director of Project Stargate, this was quite an endorsement. Assuming the capacitor current vector is pointed downward, the resulting EK force will be upwards as diagrammed in Fig. 9. The craft was reported to use about one million volts of pulsed current. A lot can be learned from studying the intricacies of this advanced design, including the use of a distributor cap style of pulse discharge and multiple symmetric, radial plates with dielectrics in between. (See reference 27 for Mark's details.) It also remains in my 'gray basket' as possibly true.

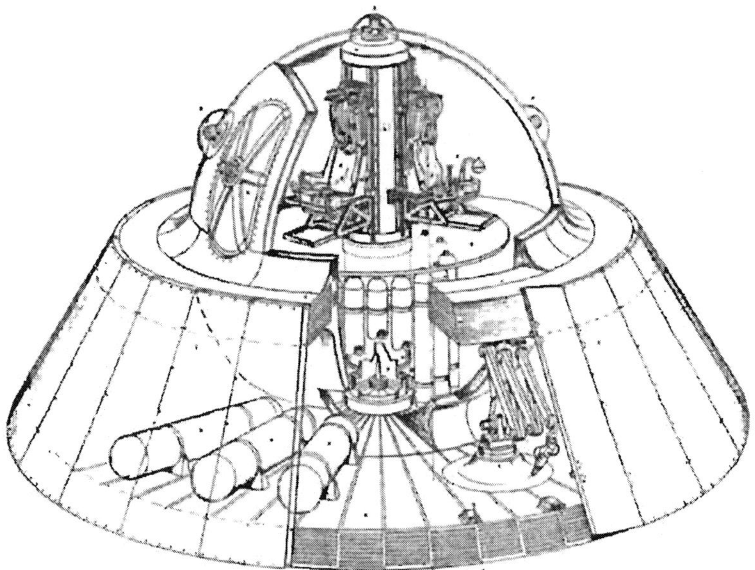

Figure 9. Electrogravitic Craft Demonstration Unit (Norton AFB, 1988) - courtesy of Mark McCandlish

Today, we still use World War II technology on land and in space. My sincere hope is that the validating science contained in *Electrogravitics II* will accelerate the civilian adaptation of this propulsion technology.

References

[1] Valone, Thomas, *Electrogravitics Systems Volume I: Reports on a New Propulsion Methodology*, 6th edition, Integrity Research Institute, Maryland, 2008, ISBN 978-0-9641070-0-7. http://www.integrityresearchinstitute.org/electrogravitics.html

[2] Loder, T., "Outside the Box Space Propulsion and Energy Technology for the 21st Century" AIAA-2002-1131

[3] Valone, Thomas, *Electrogravitics II: Validating Reports on a New Propulsion Methodology*, 3rd edition, 2008, p. 71. URL at http://www.amazon.com/s/ref=nb_ss_gw?url=search-alias%3Daps&field-keywords=electrogravitics+

[4] Zinsser, R.G. "Mechanical Energy from Anisotropic Gravitational Fields" First Int'l Symp. on Non-Conventional Energy Tech. (FISONCET), Toronto, 1981. Proceedings available from PACE, 100 Bronson Ave #1001, Ottawa, Ontario K1R 6G8

[5] Valone, Thomas *The Zinsser Effect: Cumulative Electrogravity Invention of Rudolf G. Zinsser*, Integrity Research Institute, 2005, 130 pages, ISBN 0-9641070-2-3

[6] Cravens, D.L. "Electric Propulsion/Antigravity" *Electric Spacecraft Journal*, Issue 13, 1994, p. 30 http://www.electricspacecraft.com/journal.htm

[7] Peschka, W., "Kinetobaric Effect as Possible Basis for a New Propulsion Principle," *Raumfahrt-Forschung*, Feb, 1974. Translated version appears in *Infinite Energy*, Issue 22, 1998, p. 52 http://www.infinite-energy.com and in *The Zinsser Effect* book.

[8] Valone, Thomas, "Inertial Propulsion: Concept and Experiment, Part 1" Proc. of Inter. Energy Conver. Eng. Conf., 1993, Available as IRI Report #608.

[9] See "Pulsed Electromagnetic Field Health Effects" IRI Report #418 and *Bioelectromagnetic Healing: A Rationale for Its Use* ISBN 978-0-9641070-5-2 book by this author, which explain the beneficial therapy which PEMFs produce on biological cells.

[10] Mark McCandlish's Testimony (p. 131 of *Electrogravitics II*) shows that the Air Force took note in that the electrogravitic demonstration craft shown at Norton AFB in 1988 had a rotating distributor for electrically pulsing sections of multiply-layered dielectric and metal plate pie-shaped sections with high voltage discharges.

[11] See Saxl patent #3,357,253 "Device and Method for Measuring Gravitational and Other Forces" which uses +/- 5000 volts.

[12] Saxl, E.J., "An Electrically Charged Torque Pendulum" *Nature*, July 11, 1964, p. 136

[13] Saxl & Allen, "Observations with a Massive Electrified Torsion Pendulum: Gravity Measurements During Eclipse," IRI Report #702.(Note: 2.2 lb = 1 kg)

[14] Graph of Fig. 1 from online report, Woodward and Mahood, "Mach's Principle, Mass Fluctuations, and Rapid Spacetime Transport," California State University Fullerton, Fullerton CA 92634

[15] Cramer et al., "Tests of Mach's Principle with a Mechanical Oscillator" AIAA-2001-3908 email: cramer@phys.washington.edu

[16] Woodward, James F. "A New Experimental Approach to Mach's Principle and Relativistic Gravitation, *Found. of Phys. Letters*, V. 3, No. 5, 1990, p. 497

[17] Compare Fig. 1 graph to Brown's ONR graph on P.117 of Volume I

[18] Nordtvedt, K. *Inter. Journal of Theoretical Physics*, V. 27, 1988, p. 1395

[19] Mahood, Thomas "Propellantless Propulsion: Recent Experimental Results Exploiting Transient Mass Modification" Proc. of STAIF, 1999, CP458, p. 1014 (Also see Mahood Master's Thesis www.serve.com/mahood/thesis.pdf)

[20] For comparison, 1 Newton = 0.225 pounds

[21] Zinsser, FISONCET, Toronto, 1981, p. 298

[22] Woodward, James "Flux Capacitors and the Origin of Inertia" *Foundations of Physics*, V. 34, 2004, p. 1475. Also see "Tweaking Flux Capacitors" *Proc. of STAIF*, 2005

[23] Jefimenko, Oleg *Causality, Electromagnetic Induction and Gravitation*, Electret Scientific Co., POB 4132, Star City, WV 26504, p. 29

[24] Jefimenko, p. 31

[25] Jefimenko, p. 47

[26] Brown's second patent #2,949,550 (see Patent Section: two electrokinetic saucers on a maypole) has movement toward the positive charge, so the same electrokinetic theory explained above works for both.

[27] McCandlish, Mark, "Testimony of Mr. Mark McCandlish, December 2000," *Electrogravitics II*, Integrity Research Institute, 2005, p. 131

[28] Einstein and Laub, *Annalen der Physik*, V. 26, 1908, p.533 and p. 541 – two articles on the subject of a moving capacitor with a "dielectric body of considerable permeability." Specific equations are derived predicting the resulting EM fields. Translated articles are reprinted in *The Homopolar Handbook* by this author (p. 122-136). Also see Clark's dielectric homopolar generator patent #6,051,905.

[29] Commentary to Eq. 2 states an electrokinetic impulse is produced when the "current is switched on," which implies a very steep leading edge of the current slope.

[30] See the Taylor and Schlicher patents in the Patent Section. – Ed note

[31] Ning Li was the Chair of the 2003 Gravitational Wave Conference. The *CD Proceedings* of the papers is available from http://www.IntegrityResearchInstitute.org

[32] Valone, Thomas, *The Homopolar Handbook: A Definitive Guide to Faraday Disk and N-Machine Technologies*, Integrity Research Institute, Third Edition, 2001. ISBN 0-9641070-1-5 http://www.amazon.com/s/ref=nb_ss_gw?url=search-alias%3Daps&field-keywords=homopolar+handbook

[33] Graham and Lahoz, "Observation of Static Electromagnetic Angular Momentum in vacuo" *Nature*, V. 285, May 15, 1980, p. 129

[34] See the authoritative book by Dr. Steven Greer, *Disclosure: Military and Government Witnesses Reveal the Greatest Secretes in Modern History*, Crossing Point, 2001. It provides the testimony of each witness who participated in the event, plus many more.

T.T. Brown and the Bahnson Lab Experiments

by Charles A. Yost

The Bahnson family, originally from Denmark, settled in the southern United States in the late 1700s. Several of the family members were ministers. The Bahnson Company was the first to develop and meet the need for "air conditioning" in the early factories and mills there. The company became well established in this field; it has since been sold and the name changed.

While Agnew Bahnson, Jr. was very involved with the company that had been founded by his father, he also had an overpowering interest in the possibilities of space travel, anti-gravity, and electric propulsion. Bahnson was driven by a vision of space travel to the stars.

In early November, 1957, he acquired the help of Townsend T. Brown as a consultant. Under the auspices of the Bahnson Company, Bahnson, Brown and J. Frank King began an intense series of high voltage experiments whose purpose was to find out if electric fields could produce anti-gravity propulsion.

J. Frank King was close to Bahnson. He had a strong engineering background and a similar interest in anti-gravity ideas. An entire book could be written on the efforts of these three men during the few years of their combined research.

A review of the lab notes of T.T. Brown and Agnew Bahnson reveals no clear findings of anti-gravity effects in the experiments they performed. These experiments continued at least until September of 1962. They attributed most of the force effects to ion accelerations or coulomb forces on nearby objects. Those movements and forces that were not clearly explained remained in an area of uncertain resolution because of the difficulty in controlling high voltage conditions.

While it appears that anti-gravity effects were not unveiled, this experimental research is well worth relating because it reveals efforts and techniques to produce electric propulsion by very capable individuals. They had strong resources and personal connections with some of the most respected scientists in the western world. They were seeking to learn, and it is helpful to know what they tried.

Bahnson was driven by a vision of space travel to the stars. In ...1957, he acquired the help of Townsend T. Brown as a consultant.

Townsend T. Brown remained in friendly contact with Agnew Bahnson and J. Frank King even after the experiments ended. Perhaps as a fitting summary of the research, Bahnson wrote a book entitled *The Stars Are Too High*, in which he suggested the need for certain improvements in the social condition of mankind. Bahnson died before his time in a 1964 airplane crash.

The Notebooks

The personal research notebooks of Townsend T. Brown, Volumes 1, 2, and 4, were published by W.L. Moore Publications and Research (1) in 1986. Volume 3 is unpublished, apparently because of legal problems. Brown's notebooks begin on October 1, 1955 and end on February 10, 1977. (See Note, page 12.)

But there are five other notebooks involving T.T. Brown— those of Agnew Bahnson, Jr. Written by Bahnson during the period from November 1, 1957 to September 6, 1962, they contain detailed accounts of the experiments done by the Bahnson, King, and Brown team. Notebook #4 is missing from the series.

ESJ has acquired persmission to describe technical details from the Bahnson Lab notebooks so that they can be shared with other experimenters in the search for anti-gravity or electric propulsion. It is hoped that in time all the notes will be published so that the efforts of these scientists become part of the knowledge base.

The Bahnson Lab experiments started with simple high voltage, direct current experiments. But as experiments progressed, many combinations were tried using Tesla coil RF, pulsed voltages, and electrostatics, as well as contra-rotating charged disks in electromagnetic fields. Many famous scientific names were involved in these efforts.

In the end, the sought-for results were not isolated. The experiments are nevertheless valuable to those continuing the quest of Bahnson, King, and Brown.

The Notes of Agnew Bahnson

The notes (2) start in Book #1 on November 1, 1957. It is evident that experiments by Bahnson and King had been in progress for some time. The powerful "Beta" supply (a high voltage DC transformer), the high vacuum bell jar, and other experimental set-ups were in place. A Brush transducer and tension analyzer (recorder) were used to measure forces that might be produced by a capacitor consisting of two 6 inch diameter plates spaced $2^{3}/_{16}$ inches apart, in air. Electric wind and coulomb forces predominated and masked any other kind of condenser movement.

Experiments with 3 inch diameter plates spaced $2\ ^{3}/_{16}$ inches apart were tried in a $.7 \times 10^{-5}$ vacuum. Wind effects were greatly reduced, but coulomb forces were still a big problem. (Vacuum units are not given but we think they are in mmHg from notes in other places.)

Bahnson noted that "TT Brown arrived back from Florida on November 5, 1957," and additional tests were run in the vacuum of the bell jar with voltages from the "Beta" at about 100 kv. At this voltage, the 3" diameter aluminum condenser plate pulled out of its press fit to an aluminum tube. (Murphy has a special place in his heart for experimenters.)

The condenser design shown in Figure 1 above was tested in the vacuum jar to demonstrate the Biefeld-Brown effect. First trials failed because of high voltage arcing. A neoprene seal on the bell jar was removed since it was thought to prevent the attainment of high vacuum. Nevertheless, Brown demonstrated the movement of the condenser rig, (Figure 2a, below), in air two days later. A low voltage was surged. The actual voltage was not stated, but the rig had been suspended by two wires. In addition, a ½" plexiglass rod was tried, with similar movement, Figure 2(b).

A metal plate on one end of the plexiglass rod enhanced the movement, Figure 2(c).

Details of arrangement and suspension in these experiments are not clear since no diagram ac-

Figure 1 High Vacuum Test (Nov. 9, 1957)

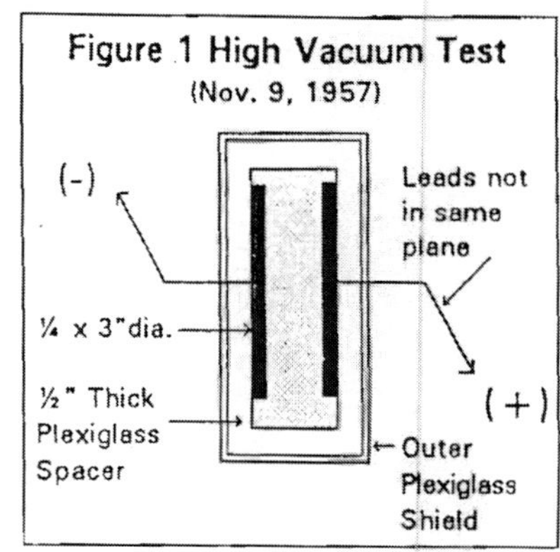

Figure 2 Air Suspension Tests (Nov. 11, 1957)

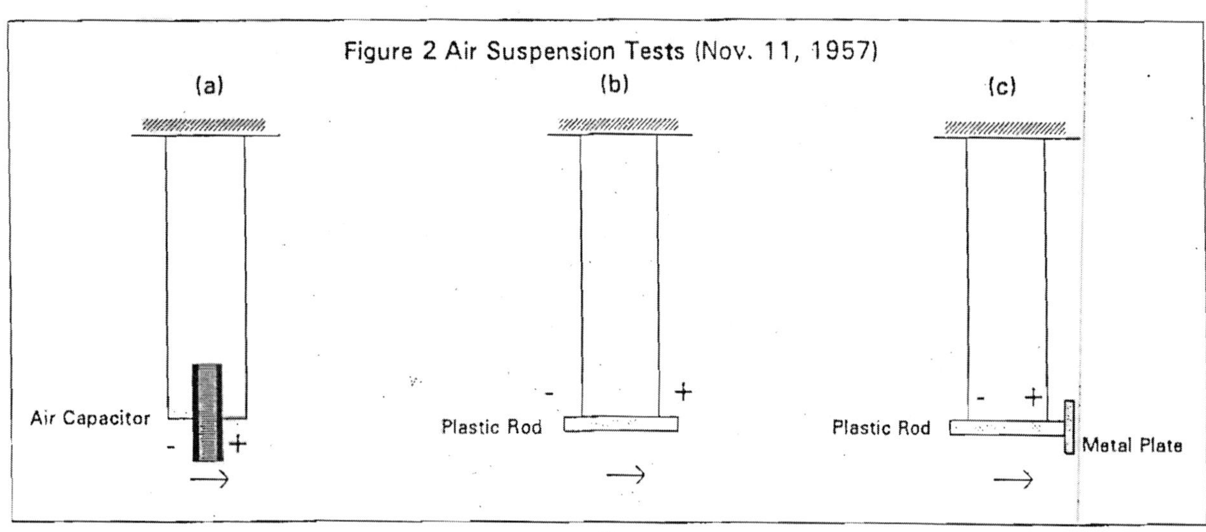

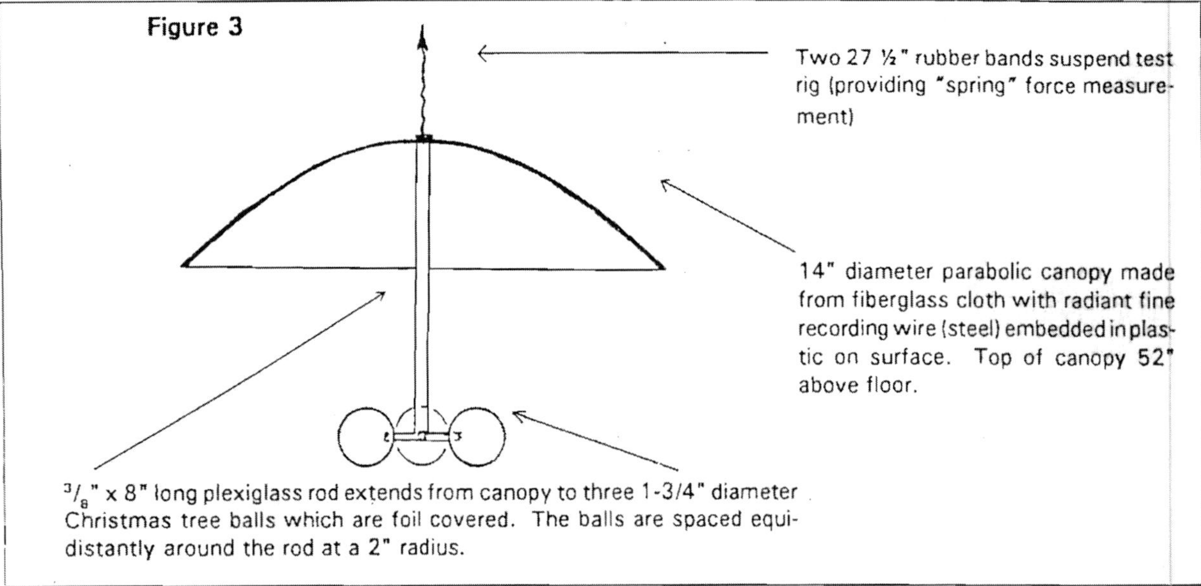

companies the description. We assume the setup was something like that shown in Figure 2(a), (b) and (c).

They concluded that voltage pulsation and field shaping caused coronal leakage into the air, which created movement forces.

Field Shaping Canopy

On December 21, 1957, Bahnson's notes describe a new test rig design as shown in Figure 3.

At 100 kv the rig draws 2 milliamps steady state, and upon a sudden charging pulse it rises about three inches before arcing.

There is a tendency for the rig to rock and rotate clockwise under steady state voltage. (The (+) load is thought to come in from the top on this rig as it does in other similar tests, although this is not specifically mentioned in the notes.)

On December 26, 1957, the rig weight is given as 170 grams. The "lift" was 26 grams at 150 kv and 3.6 milliamps.

On December 27, 1957, Bahnson commented: "...varying the distance (focus) of the upper curved 'plate' (+) [canopy] and the lower 'plate' [balls, plates, etc.] will vary the volts to amps relationship. With the plates close together we seem to get more amps per volt and vice versa."

RF Pulsing Suggested

On January 5, 1958, Bahnson suggested to Brown the idea of RF (radio frequency) pulsing of the test apparatus, thinking such a technique might trigger latent energy in the existing static field with a minimum of power. The idea was to input an RF into the "lift" canopy so that no current flowed until the grid was "tuned" to a resonant frequency. (See Figure 4.)

The resonant frequency would relate to the spacing between the (+) canopy and (-) cathode ball. It was hypothesized that the energy of the electrostatic field was stored in the space surrounding the rig in an "aura" or "ether-like" reservoir, with a release at the resonant frequency—a kind of flying triode battery.

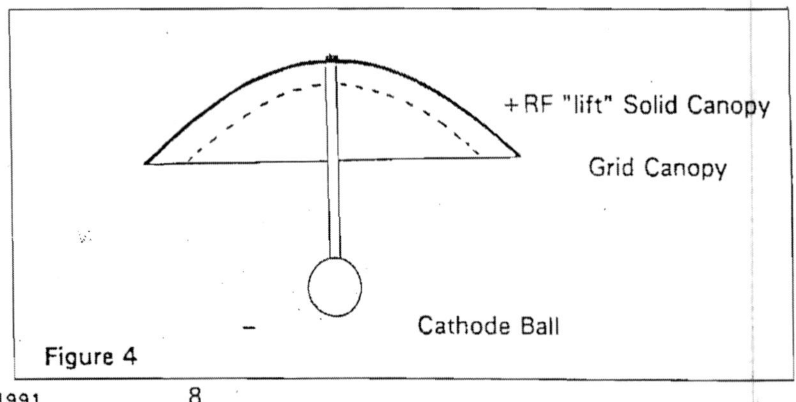

Figure 4

Variations of the Parabolic Canopy

By January 16, 1958, the Bahnson experiments were concentrating on variations of the parabolic canopy and its spacing from the suspended ball or grid below. No mention of the RF pulse or resonant type of experiment was made. A change to the canopy having a (-) polarity and a (+) grid 10 inches below resulted in a 110 gram lift force at 100 kv and 2.7 milliamps. Figure 5 illustrates the configuration.

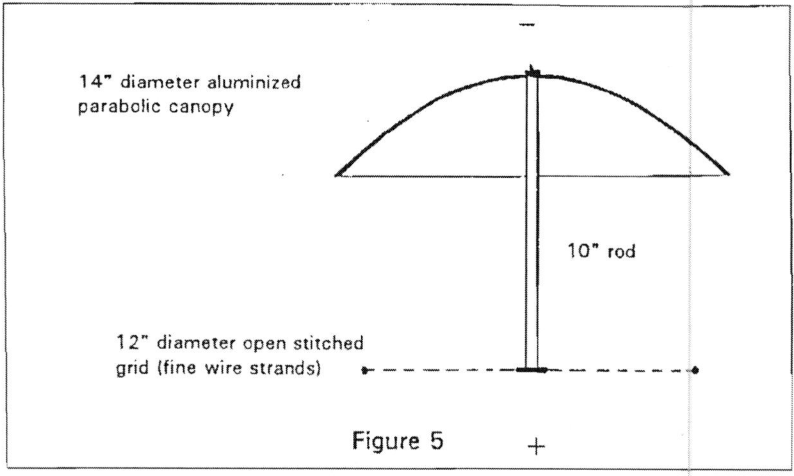

Figure 5

Lift Efficiency

On January 17, 1958, the question of lift efficiency came up. Favorable performance of the canopy configurations were taken to be 60 grams lift at 164 kv and 1 milliamp. The "work efficiency" could not be resolved because neither rig velocity data nor wind velocity data was measured.

The efficiency question was not resolved. However, we can assume a "wind" and analyze what kind of "wind" might be required to accomplish the 60 gram lift force of the 14 inch diameter rig. Assuming sea level atmospheric conditions, the average pressure required to lift 60 grams would be the lift force (60 grams) divided by the 14" diameter canopy area.

Lift = 60 grams = .13 lb

$$\text{Area} = \pi D^2/4 = \frac{\pi(14/12)^2}{4} = 1.07 \text{ Ft}^2$$

$$\text{Pressure} = .13 \text{ lb}/1.07\text{Ft}^2 = .12 \text{ lb/Ft}^2$$

Given the sea level air density (ρ) to be .00238 slugs / ft^3, the air velocity required to produce this pressure would be:

$$V = (2P/\rho)^{1/2} = (100.8)^{1/2} = 10 \text{ ft/sec}$$

Where's the Wind?

The curious thing so far in the notes is that there has been no mention of "wind" from the canopy rig. A 10 ft/sec wind is quite obvious and should have been noticed.

The high voltage power supply output is estimated at 164 watts (164 kv x 1 milliamp), or nearly ¼ horsepower. Less than 1% of this power would be required to produce the 10 ft/sec wind. Perhaps there was a lot more wind during electrical pulse; or perhaps the electrical energy applied itself in other ways.

Evolution of Canopy Shape

By January 23, 1958, the canopy shapes had evolved to the concept shown in Figure 6.

No significant discussion appeared to come of this configuration.

A Thought

On January 28, 1958, mention is made of a concept that utilized two contra-rotating masses, charged and under the influence of an electromagnetic field. It

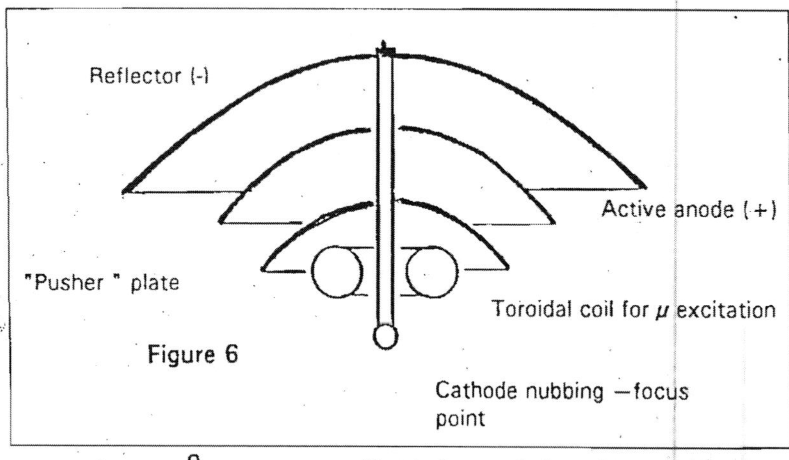

Figure 6

was suggested that this may alter the gravitational field if space consists of an "ether-like" medium.

Pendulum Period

In June and July of 1958, a series of pendulum experiments were conducted to determine the effects of high voltage on the pendulum swing period.

The formula for the simple pendulum period (T) is:

$$T = 2\pi (L/g)^{1/2} \qquad (1)$$

The Bahnson experiments indicated that the pendulum period was modified by high voltage as expressed in the following equation:

$$T = 2\pi \sqrt{\frac{L}{(g - KE^2/m)}} \qquad (2)$$

where:

T = pendulum period (sec)
L = pendulum length (ft)
m = pendulum mass (slugs)
E = voltage (volts)

K = empirical constant

g = acceleration of gravity (ft/sec^2)

From this equation it is evident that as the voltage is increased, the (KE^2/m) term subtracts from (g) and the period increases. It suggests that if the voltage were high enough, (g) would be entirely negated. However, the range was not broad enough to draw such a conclusion. In addition, other experiments indicated that pulsed voltages must also be considered.

Wind Velocity

On July 18, 1958, it was noted that the wind velocity was not purely electrostatic. A year later on July 2, 1959, Bahnson tabulated wind velocities at various voltages as shown below in Figure 7.

A special setup, like a venturi duct, was used to produce this data and so no direct correlation is made to lift weight. However, the wind velocities are quite low and as such could not account for all the lift measured in earlier tests.

My notes at this point were rather sparse and so details of the setups are not described.

Townsend T. Brown:
Prelude to Joining Bahnson

It seems worthwhile to review T.T. Brown's activity just prior to his joining the Bahnson Lab effort. From 1955 to 1957, Brown resided in Leesberg, Virginia. Before going to Bahnson Labs, he returned briefly to his residence in Umatillo, Florida.

In 1955, Brown concerned himself with the lighter and heavier isotopes of particular elements, particularly the rare earths. He thought the lighter isotopes ought to exhibit a spontaneous and high evolution of heat. He did not

Figure 7. Wind Velocities at Various Voltages

KV	Air Exit Vel(ft/sec)	Milliamps Line	Milliamps Panel
20	0.00	0.00	0.00
40	0.00	0.00	0.00
60	0.38	0.03	0.07
80	0.42	0.10	0.16
100	0.75	0.15	0.26
120	1.00	0.25	0.43
140	1.25	0.35	0.65
160	1.53	0.50	0.95
180	1.76	0.70	1.32
200	3.33	0.95	1.75
210	3.50	1.05	2+

mention the conditions required, but indicated that up to that date such an effect had not been discovered. The heavier isotopes, he thought, should absorb heat.

Brown mentioned that C.T. Brush performed an experiment in which he produced super-light (weight)hydrogen, possibly created by some sort of preferential selection of ions in water or during the electrolysis of water.

I have interpeted some of Brown's terms in Figure 8 below.

Increased or decreased electrical charge also included the transient process of the charge being increased or decreased. To test his hypothesis, Brown suggested setups whereby samples could be hit by lightning!

By January of 1956, Brown was developing a hypothesis that "mass was somehow composed of two parts": a gravitational mass (Mg) which reacted to gravitational fields; and an inertial mass (Mi) which reacted to acceleration. He considered these to normally be equal in quantity normally, but that variation of the ratio would show up as an atomic isotope which was lighter or heavier.

By mid-February, his hypothesis had led to a concept of an "Inertial Differential Electrogravitic Motor." This had further evolved within the month to a rotating disk shape with a net forward thrust.He equated this to the ion pulse effect in the force developed by a simple capacitor in a vacuum. Brown was still trying to explain the Biefeld-Brown effect he had discovered many years earlier.

He thought the directional force developed by the electric dipole (capacitor) varied with the rate-of-change of the voltage. This force, independent of any ion or coulomb mechanics, operated in the negative to positive direction as the voltage increased.

The direction of force was presumed to reverse if the voltage were decreasing. Brown had in mind the use of selected isotopes in order to get the maximum effect.

On April 7, 1957, Brown made an interesting statement: "The vacuum spark is apparently not due to a flow of electrons, although a flow of electrons may accompany the discharge." This provides interesting food for thought. The rate of change caused by the spark became part and parcel of the electro-gravitic mechanism.

By September, Brown was examining the possibility of measuring the change in weight of "excited" materials (over months of time). "Excited" materials could be produced by friction, pulverization, or impact. This led to further ideas on dipoles within dielectrics being sensitive to natural external field changes, with the dielectrics literally recording these field changes over time. By November, 1957, Brown had embarked on a path which later led to his extensive research into "rock electricity" or "petro-electricity."

It was at this period in Brown's life that Bahnson made contact and requested Brown's assistance in electric propulsion and anti-gravity research.

Brown summarized his thoughts at this time with a sketch, shown

Figure 8

Gravitational Isotope: The specific gravity of a mass as measured in the Earth's gravity field

Indo-gravitic: A mass taking on more positive charge has:

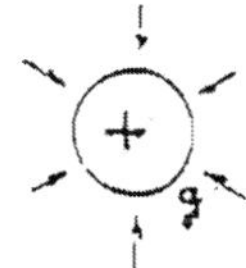

- increased gravity attraction
- increased inertial mass
- decreased gravity mass (weight)

Exo-gravitic: A mass taking on more negative charge has:

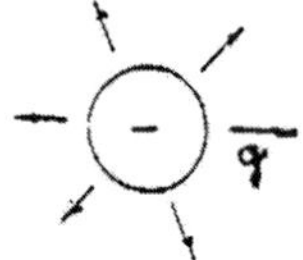

- decreased gravity attraction
- decreased inertial mass
- increased gravity mass (weight)

in **Figure 9**, of an experiment to measure the weight loss of a weight suspended in a vacuum when bombarded by positive ions.

Hereafter, his thoughts shifted to the research activities at the Bahnson Labs. 

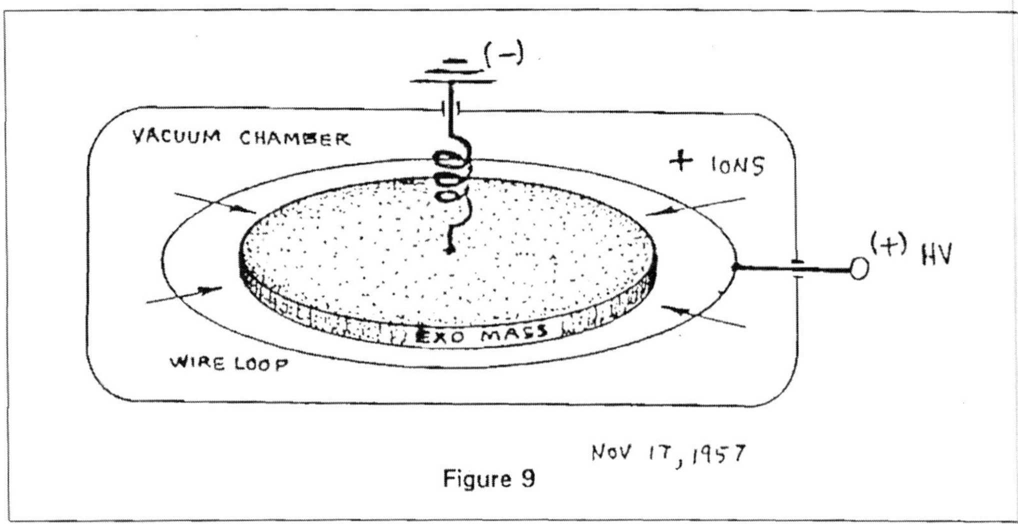

Figure 9

References

(1) *T. T. Brown Notes,* W.L. Moore Publications and Research, 4219 W. Olive St., Suite 247, Burbank, CA 91505

(2) Unpublished notebooks of Agnew Bahnson, Jr.

Note

ESJ has tried to contact Mr. Moore by telephone and by mail without response. A recent letter from **Klaus Schlecht** in Germany indicates that Moore Publications is now owned by *Peregrine Journal* and only Volumes 1, 2 and 4 out of a six volume set were ever made available. A portion of Mr. Schlecht's letter is printed here.

> ...With the delivery of vol. 4, Mister Moore wrote to me, that vol. 4 was sooner ready than the outstanding vol. 3, 5 and 6, which he intented to send me as soon as possible after their print readiness. Although I have remembered him several times for the delay of the outstanding volumes, I heard nothing more in that case from him. At 12.1.1991, the Peregrine Communications wrote to me, that the W.M.L. Moore Publications belongs now to them, and after their records, I should have a credit, which could be balanced by an equivalent delivery of the *Peregrine Journal.*
>
> Of course I was not interested in the *Peregrine Journal.* Neither they send me my money back (half amount), nor took they every responsibility for the delay of the missing T.T. Brown Notebooks, which have an unmeasureable value for me....

§ *To be continued* §

History

T. Townsend Brown Notebooks

Charles A. Yost

***Charles Yost** offers excerpts from the 1955 to 1977 lab notebooks of T.Townsend Brown (1905-1985).*

Yost can be contacted at ESJ, 73 Sunlight Drive, Leicester, NC 28748.

Volume I–1955

October 1, p. 1 Leesburg, Virginia
Brown begins his first notebook, mentioning that all too often he has lost valuable information because he did not keep a notebook. These first notes relate to gravitation and electrodynamics.

October 7, pp. 3+
A note that Glen L. Martin and General Electric will be part of the space program. He mentions K. Jessup's "In the Case for the UFO" and refers to the ability of ancient peoples to move enormous stones. Brown mentions gravitational isotopes and suggests that in these early times (about 200,000 years ago), things may have simply been lighter in weight. [This is analogous to the idea expressed by Stan Deyo, that light used to travel much faster.]

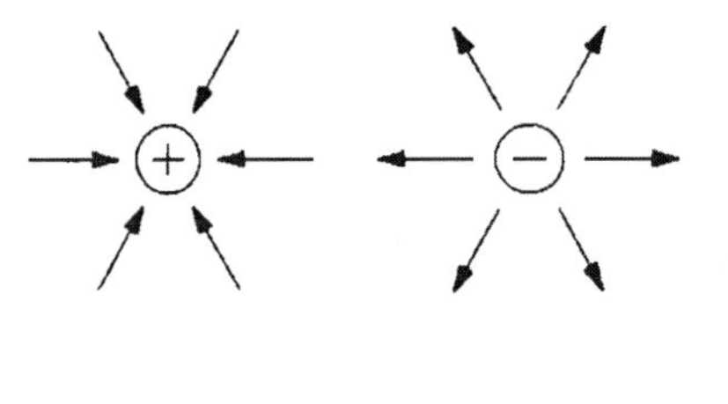

Fig. 1 Brown's concept.

p. 13
Brown's Concept (Fig. 1)
increased negativity (exogravitic field) causes:
- increase of gravitational mass (weight)
- decrease of inertial mass
- decrease of gravitational attraction

increased positivity (endogravitic field) causes:
- decrease of gravitational mass (weight)
- increase of inertial mass
- increase of gravitational attraction

December 25, p. 18
Brown suggets that gravitational potential is synonymous with the potential of the negative effluvium (Fig. 2).

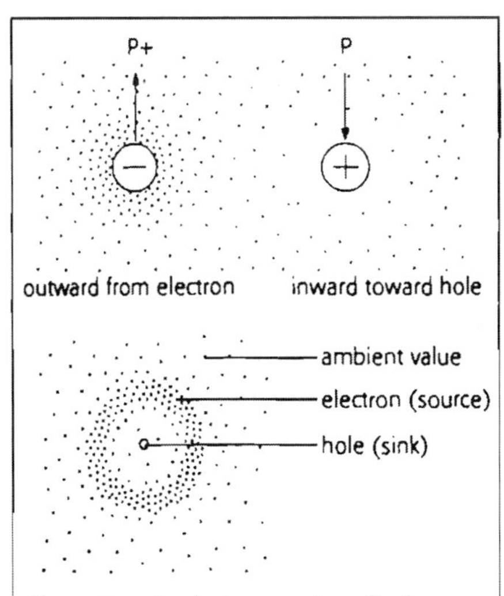

Fig. 2 The ether is the negative effluvium—the "elastic" compression of potential energy.

1956

January 7, p. 21 Charles Brush experiments
Heat treating and working with metals causes metals to lose weight. Initial compression, hammering, etc., increases the specific gravity of metals, but continuing such action decreases the specific gravity. Might also the tensile strength (for example, the ability of a piece of metal to withstand continuous flexing before it breaks) decrease at the same time? .

February 5, p. 37
Three factors affect the rate of fall (acceleration) of a freely falling body:

1. the intensity of the gravitational field
2. the susceptibility of the material being acted upon to the field
3. the inertial mass of the material.

p. 38
In a gravitational field,
acceleration = $m_g\ f_g/m_i$
f_g = gravity gradient force
m_g = gravitational (susceptibility) mass
m_i = inertial mass

Comparing m_g to m_i , or g to i,
$g/i = 1 \Rightarrow$ normal mass-weight equivalence
$g/i > 1 \Rightarrow$ heavy gravitational isotopes prevail
$g/i < 1 \Rightarrow$ lighter gravitational isotopes prevail

Brown suggests the use of a centrifuge (inertial acceleration) to determine if a "gravity" mass differs from an inertial mass.

April 7, p. 75
In a vacuum (10^{-6}mm Hg or less)–a simple vacuum capacitor will appear to flash as the voltage increases. Concurrent with the vacuum spark, an impulse force is notable in the direction from negative to positive.

p. 77
Successively higher voltages, starting at 30-40kV, and going up to 150kV, must be applied to allow flashing spark-over to continue. The anode takes on a reddish glow upon flashing and the cathode has bright starlike points. Impulse forces are on the order of thousands of dynes.

September 9, p. 83
Brown reiterates that grinding or pulverizing sand or rock into sand might cause it to lose weight. Friction or irradiation by sunlight while grinding are believed to supply the necessary energy.

1957

November 5
[Brown and Bahnson connect.]

Dec. 27, p. 98 Umatilla, FL
Brown discusses field shaping using shaped dielectrics (Fig. 3).

1958

January 1, p. 102 Winston Salem, NC
In the first notebook entry directly referring to the Bahnson experiments in Winston-Salem, NC,

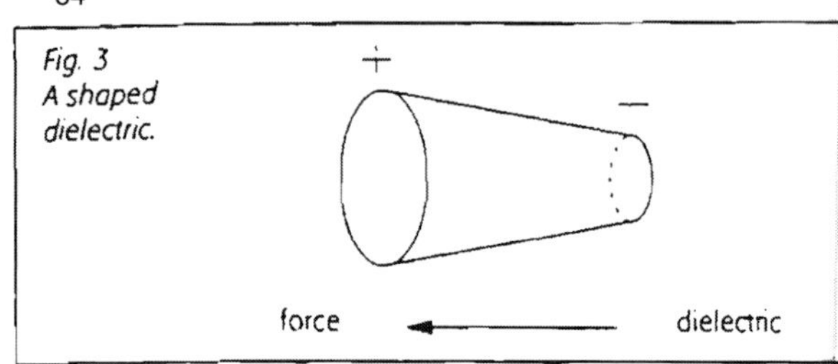

Fig. 3 A shaped dielectric.

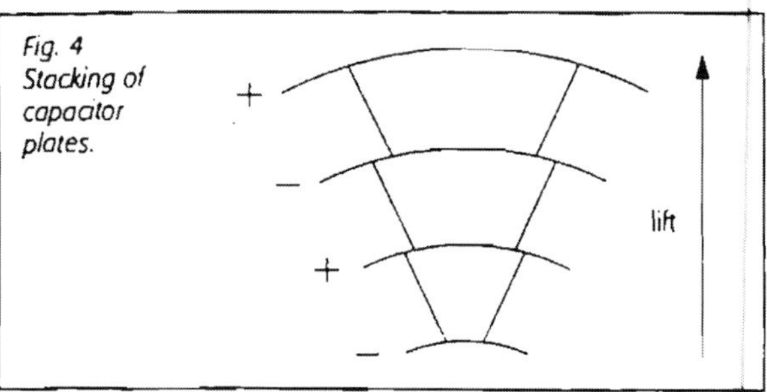

Fig. 4 Stacking of capacitor plates.

Brown notes the stacking of capacitor plates for field shaping (Fig. 4).

January 5, pp. 103+
Brown describes possible details of the Adamski Venusian spaceship, reconstructing Adamski photographs and pointing out similarities to their (Bahnson/Brown) own models. The interpretation follows the lines of Brown's previous documents with thoughts of gravitic isotopes, aluminum silicates and rare earths in the field of a capacitor.

p. 108
Brown speculates on the cause of red coloration reported with "saucer" acceleration.

January 12, p. 111
Brown continues speculating on the "Venusian Scout Ship" design. The three balls are assumed not to rotate but rather to be used for setting flight direction and counteracting precession forces caused by a rotating toroidal coil (power coil).

January 19, p. 119
[Fig. 5.]

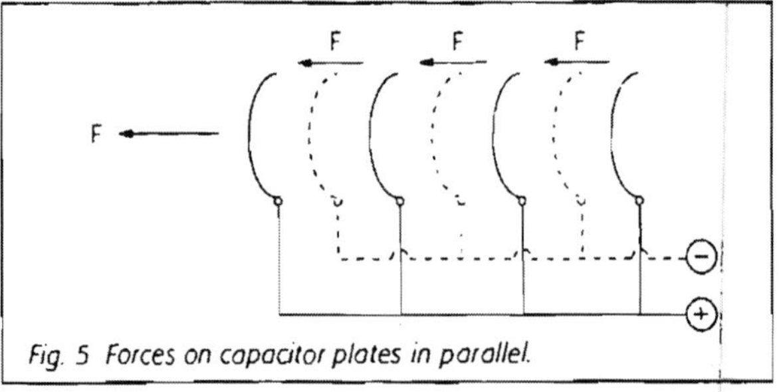

Fig. 5 Forces on capacitor plates in parallel.

p. 120
[Fig. 6.]

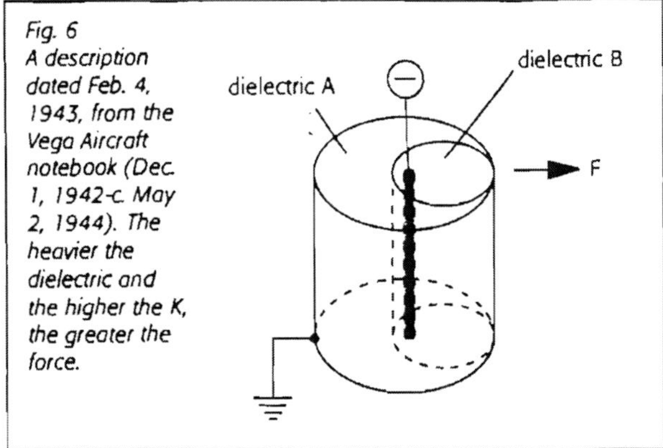

Fig. 6
A description dated Feb. 4, 1943, from the Vega Aircraft notebook (Dec. 1, 1942-c. May 2, 1944). The heavier the dielectric and the higher the K, the greater the force.

March 25, p. 124
J. Frank King reviews Brown's notes. Brown is still leaning toward the Adamski design. An RF inductor is introduced along with DC high voltage and a three-element electrode system.

April 7, p. 132 Walkertown, NC
[Fig. 7.]

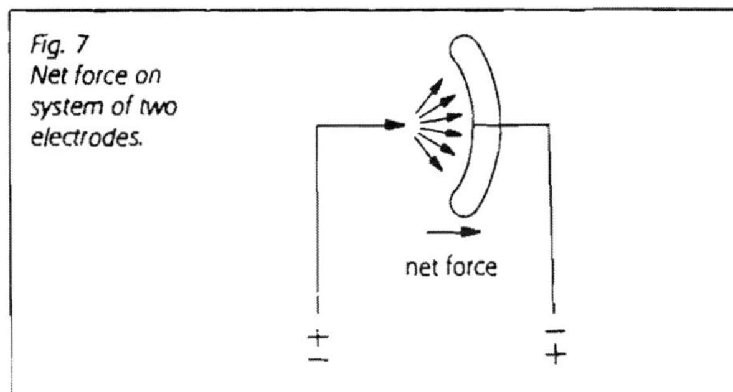

Fig. 7
Net force on system of two electrodes.

May 1, p. 148 Winston Salem, NC
The hydrostatic pressure developed by the oil within an insulating tube may be used to detect the force within a dielectric medium (Fig. 8).

Volume II—1967

October 23, pp. 21+ Santa Monica, CA
Brown notes, "During the period from October 1958 to October 1967 (nine years) no notes were made."

Brown writes about the construction of loudspeakers and air movement using high-voltage supplies. [The relationship with Agnew Bahnson, Jr. ends around 1965 with Bahnson's death. However, Brown continues to correspond with J. Frank King.]

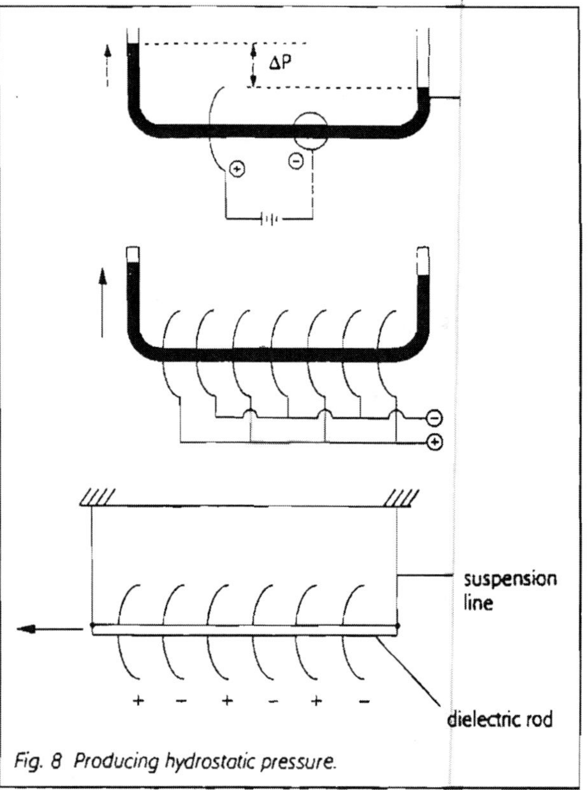

Fig. 8 Producing hydrostatic pressure.

1970

May 31, p. 25 Palo Alto, CA
Brown begins investigating the use of capacitors to detect gravitational waves. He starts with four 1 μF, 25kV capacitors producing loud pops and whistles. Several circuit variations are sketched.

1973

March 26, p. 27 Catalina Island, CA
A triboelectric, high-voltage generator is described in which a barium titanate (high K) rod is rotated in oil (low K). The rubbing between the two is anticipated to make a high-voltage differential with the barium titanate (+).

March 30, p. 33
Red sand from Sorrento, FL undergoes tribo-excitation when shaken in a quart jar. Before shaking, the sand weighs 14 lbs. 14½ oz.; after shaking it weighs 14 lbs. 14¼ oz. Brown is reviewing his idea that excited bodies lose weight. He considers that "sparking" to sand may produce a weight loss effect.

April 24, p. 40
Brown talks of Charles Francis Brush again and his work published in the *Physical Review*...

showing that different materials fall at different velocities. Brown is particularly fascinated with the rare earth elements and silicates (clay).

August 19, p. 62
Brown notes that Fernando Sanford measured cyclic, sidereal variations in the resistance of a wire, which could not be explained.

August 20, p. 68
Sanford observed a copper wire, 1mm dia. × 120cm long, from Feb. 17 to May 18, 1892. Its resistance variation ($6 \times 10^{-4}\ \Omega$) was cyclic under carefully controlled conditions.

September 4, p. 87
Relating to Sanford's book, *Terrestrial Electricity*, perhaps an electric field gradient surrounding the earth produces a sidereal factor of resistance change (apparent).

September 11, p. 113+
Many thoughts on the connection between gravity, capacitive storage, inductance storage and the variation of resistance with sidereal time. He considers that there is a gravity vector in the direction of electron flow. He realizes that a leakage current exists in the capacitor dielectric and that this may represent a direct conversion of electron flow to a gravity force vector. The gravity field of earth was therefore not being acted upon and the kinetic energy of the capacitor was not derived from the earth field. He relates that his early experiments showed the "gravitator" force was a function of mass. Lead monoxide as a dielectric loaded in paraffin was usually used. The conductivity of the mass was never investigated (Fig. 9).

Both field and current are now considered as necessary to create a gravity vector within the semiconductive capacitor, requiring a very high voltage. The electron flow is from negative to positive, passing through the semiconductive mass and producing a gravity vector in that direction. Such a gravitator is considered a good sensor to detect variations of electrical and gravitational waves in the space around the earth.

September 16, p. 132
Watching his μA meter measuring currents through a resistance, Brown notes a sudden jump. The cause of such jumps is not easily explained, but is part of regular diurnal variations. Brown settles into some long term serious measurements of voltage, resistance and current changes vs. time.

36

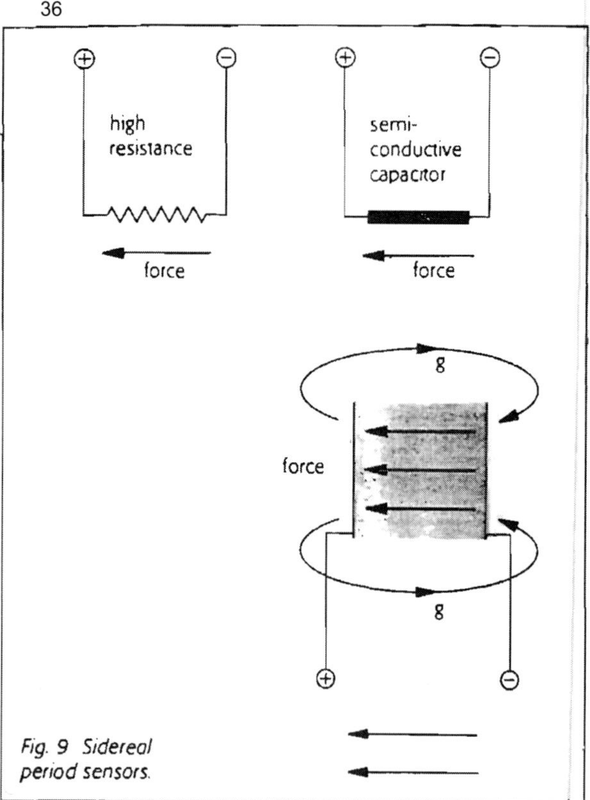

Fig. 9 Sidereal period sensors.

Volume III (missing)

Volume IV—1975

January 28, pp. 1+ Honolulu, Hawaii
Brown is in Hawaii measuring the output of his sensors (resistors, capacitors, etc.) and "rock" electrical variations. Certain rocks like granite appear polarized and do not need a diode in the test circuit (Fig. 10).

March 10, p. 36
A volcanic rock from Waikiki, 10cm in diameter, was washed and oven dried at 400°F. Copper print electrodes were painted on it after it cooled. The rock output about 60mV. When connected to the recorder it immediately showed regular pulsations about 1 second apart.

March 18, p. 42
Brown notes this is his 70th birthday.

March 24, pp. 43+
Rocks act like capacitors, they develop charge

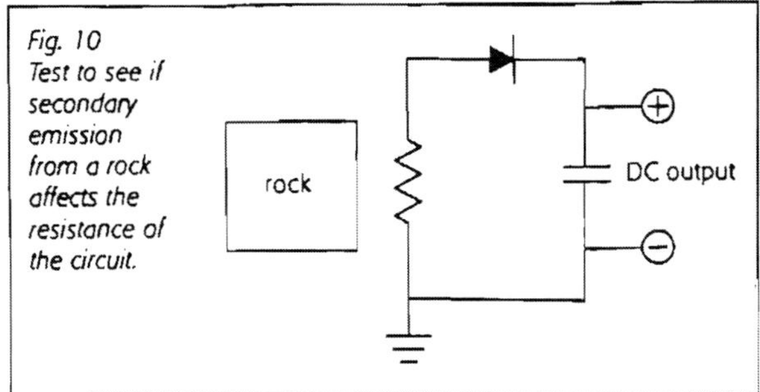

Fig. 10 Test to see if secondary emission from a rock affects the resistance of the circuit.

and keep it. Brown speculates about the pyramids at Gizeh as being electrical generators or polarizers.

1976

April 23, p. 112 Univ. of CA Berkeley & Sunnyvale
He mentions making a sensor out of a tungsten carbide and carnauba wax mixture, keeping it under a high voltage field while the wax cools and hardens. A rock hard mixture of glycerine and litharge (PbO) also makes a good sensor.

May 6, pp. 113+
Brown mentions working with Jim Jardine of UC Berkeley.

May 10, p. 115
Brown finds high capacitance electrolyte capacitors to be good sensors of diurnal variations.

May 29, pp. 223+
Speculation continues on the great pyramids.

1977

February 10, p. 151
Brown's last entry, still recording rocks and capacitor data.

[Editor's Note: T. Townsend Brown died in 1985. ESJ 19 will provide some newly acquired data from Robert Dadd, who worked with Brown in the 1970s.]

The Tesla Papers Affair

Source: Name witheld by request. The Source is from Belgrade, in the former Yugoslavia.

Documents from a European Source express concern about the future of the contents of the Tesla Library in Belgrade. The Source has reason to believe that on more than one occasion, Tesla papers from the library have been sold by private parties in authority at the museum. Other signs of corruption are also visible. The archives and catalogues of the Tesla Museum, as of April 1996, are said to have been closed to all but the highest officials, who are suspected of misusing the museum for personal advantage.

The Source reports that he has seen an original copy of a Tesla manuscript, bearing the stamp of the Tesla Museum, for sale underground. Rumors of other documents floating around for sale are not uncommon. Attempts to conduct a police investigation into the mishandling of museum documents have failed.

Political suspicions feed into the fears of persons interested in preserving the integrity of the Tesla Museum's materials. The library closed after a scandalous visit to the museum by members of the Japanese cult, Aum Shinrikyo. Patrons are also afraid that the library is being partisan to the Soviets. The fears are compounded by allegations that more than 1000 names have been removed from the catalog of correspondents for unknown reasons. Josef Stalin was one such correspondent, and a letter he wrote inviting Tesla to the Soviet Union, which was known to be kept in the Tesla Museum, is no longer there.

The Source is concerned that Tesla's documents are not safe under the current museum administration; but is worried that the political instability may cause the library to be turned over to other political cronies or appointees with no regard for the value of the documents in the museum. Late-breaking rumors from the Source indicate that the museum is further imperiled by a potential relocation to Zagreb, Croatia.

This Japanese interest in Tesla technology has been under investigation by the US Senate Permanent Subcommittee on Investigations, Committee on Governmental Affairs chaired by senators William V. Roth, Jr. and Sam Nunn. Yumiko Hiraoka, sect leader of the New York City chapter of Aum Shinrikyo, testified at the hearing on global proliferation of weapons of mass destruction on Oct. 31, 1995.

A Theory of Electrogravitics

by Paul LaViolette

General relativity fails to predict a connection between electric and gravitational fields, a fact which Einstein himself found troubling. While certain unified field theories do predict a unification of the fields, this is supposed to take place only at very high energies (in excess of tens of trillions of electron volts). However, experiments carried out by T. Townsend Brown, Paul A. Bielfeld, and others, suggest that electric and gravitational fields may be strongly connected at voltages as low as 105 ev. (1,2). If true, this would indicate that standard field theories are seriously flawed.

Subquantum Kinetics

Only one physics theory to date has predicted the existence of an electrogravitic linkage at low energies. This is subquantum kinetics.[3-5] Those interested in electrogravitic experimentation might find it useful to have a brief exposure to this novel theory.

According to subquantum kinetics (SQK), gravity potential can adopt two polarities, instead of one. Besides the matter-attracting gravity potential wells described in standard physics, it also allows the formation of matter-repelling gravity potential hills. Moreover, it predicts that these two gravity polarities should be directly matched with electrical polarity. Positively charged particles such as protons would generate gravity wells, while negatively charged particles such as electrons would generate gravity hills. Electrically neutral matter remains gravitationally attractive because the proton's G-well marginally dominates the electrons G-hill.

Consequently, according to SQK, the negative ion cloud behind Brown's disc should form a gravity potential hill while the positive ion cloud ahead of the disc should form a gravity potential well. (See *Figure 1.*) As increasing voltage is applied to the disc, its gravity hill and well become increasingly disparate and the gravity potential gradient between them increasingly steep. In Rose's terminology, the craft would find itself on the incline of a gravitation "hill." Since gravity force is known to increase in accordance with the steepness of such a gravity potential slope, increased voltage would induce

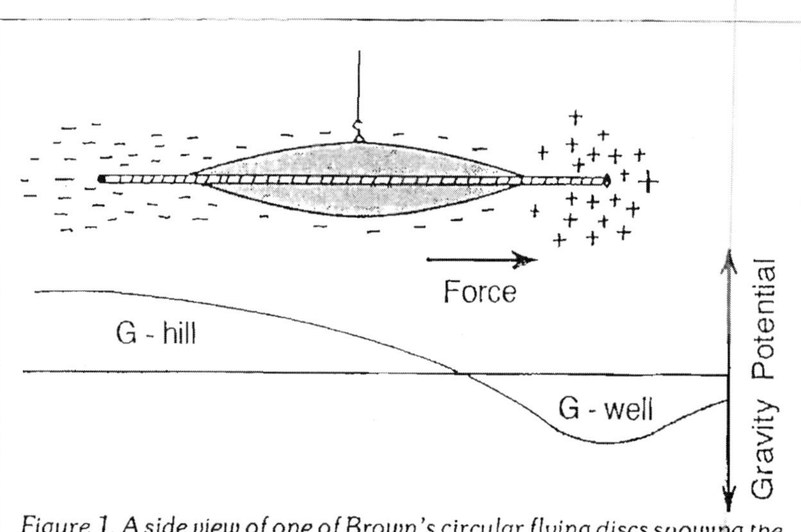

Figure 1. A side view of one of Brown's circular flying discs showing the location of its ion charges and induced gravity field.

an increasingly strong gravity force on the disc and would act in the direction of the positive ion cloud. The disc would behave as if it were being tugged forward by a very strong gravitational field emanating from an invisible planet-sized mass positioned beyond its positive pole.

To see how SQK predicts the existence of a bivalent gravitational field correlative with the electrostatic field, let us consider a few details about the theory. First, SQK presumes that space is filled by an ether that serves as the substrate for all physical form (fields, energy, and matter). However, this ether is substantially different from the mechanical ethers of classical physics in that its constituent "etherons" not only mechanically diffuse from one place to another, but also react with one another in a chemical-like fashion. This departure from traditional ether representations was not arbitrary, but was developed through an extension of the general system theoretic approach.[6] This powerful general systems methodology makes it possible to theorize in a credible fashion about the subquantum/quantum realm, a level of nature that resists direct observation. Since classical mechanical ether theories as well as many ad hoc theories in conventional physics are not grounded in the systems approach, they suffer considerably. One could argue that SQK is the only methodology to date that provides a legitimate framework for reasoning about physical phenomena.

The SQK ether is not of unitary composition, but consists of etheric particles of various specific types (A, G, X, Y, etc.). These are postulated to react with one another in a certain manner, as represented by the "Model G" reaction system shown in *Figure 2*. Model G also permits reverse reactions that occur at a rate far lower than the forward reactions, and which normally would be represented by reverse arrows (not shown). When simulated on a computer, together with assumptions about the diffusion rates of each type of etheron, the Model G reaction system produces stationary ether concentration waves that have properties similar to subatomic particles. (See *Figure 3*.) Equating ether concentration with energy potential, these emergent waves are essentially stationary energy potential scalar waves.*

*** Author's Note**

Wave phenomena not only express themselves in nature in the form of mechanical waves, but also in the form of chemical waves (i.e., reaction-diffusion waves). The latter phenomenon, unknown until the mid-twentieth century, is fundamental to SQK.

These particle-like patterns have the unique ability of generating a (1/r) potential field about themselves. A particle's G ether concentration profile is equated with its gravity potential field while its X and Y ether profiles are equated with its electric potential field. The X electric field component, not shown in *Figure 3*, mirrors the Y field component. That is, X is at a minimum at the center of a positively charged particle and at a maximum at the center of a negatively charged particle.

A particle generates its electric field as follows. The Model G reaction

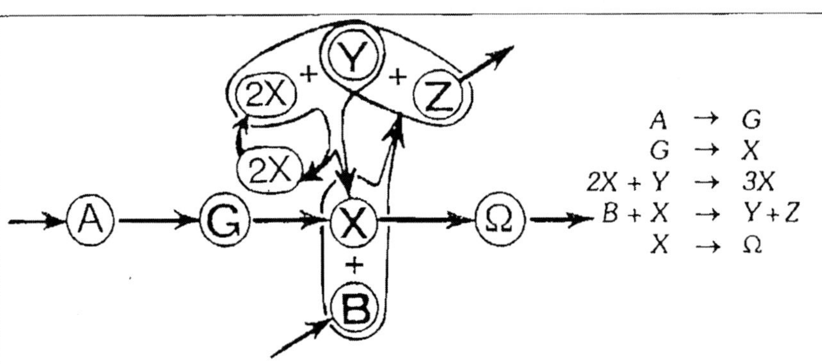

Figure 2. A representation of Model G, depicting ether reactions postulated to take place in the ether. Variations in the concentration of G, X, and Y would constitute the potential fields that generate physical form.

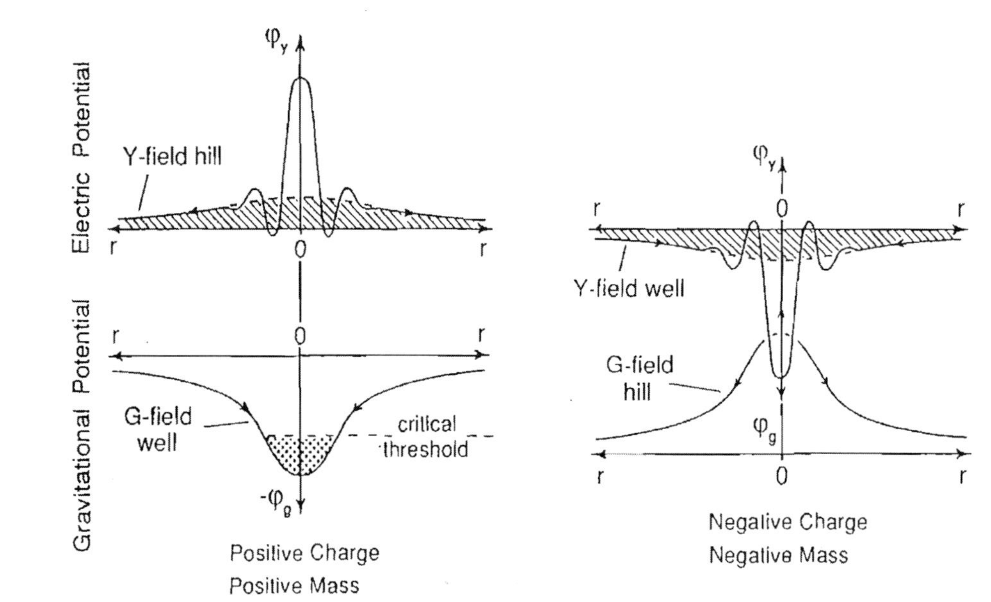

Figure 3. Energy potential field profiles for a positively charged particle (left) and its negatively charged complement (right) shown in radial cross section. The vertical axes plot the field's magnitude and the horizontal axes plot radial distance from the particle's center. Upper shading indicates the electric potential biases associated with positive or negative charge.

system requires that a proton's core produces a flux of Y etherons which diffuse radially outward to form a 1/r Y-hill. Also, it requires that the proton's core consumes a net negative flux of X etherons, hence X etherons from the environment diffuse radially inward to form an X-well that also has a 1/r form. In an electron, the X and Y fields and fluxes are just opposite to those in a proton, as are the productions and consumptions in its core.

A particle generates its gravity field as follows. The low X concentration in the proton's core is responsible for producing a low G concentration by virtue of the reverse reaction G ←X (see **Reference** 3 for more detail). The resulting net consumption of G in the proton's core induces an inward flux of G etherons and thereby produces a (1/r) G-well-- a positive mass gravity potential well. The situation is just the reverse in the electron. There, an excess G etheron production generates an outward G flux and produces a (1/r) G-hill--negative mass gravity potential hill.*

* Author's Note

To fit the standard view of gravity, a neutron should have a positive gravitational mass comparable to that of a proton-electron pair; that is, it should produce a G-well whose depth is comparable to the residual G-well produced by an electrically neutral hydrogen atom. However, since Model G predicts that electrostatic and gravitational fields are coupled, this would imply that a neutron should exhibit a very small positive electric charge, perhaps a billion times smaller than that of a proton. The presence of such a charge might be detected by carrying out sufficiently sensitive measurements.

When a voltage difference is generated between two locations, as shown in *Figure 1*, the resulting electric field is complemented by a gravity field that proceeds from a maximum G potential at the negative pole to a minimum G potential at the positive pole. This G field induces a gravitational force on intervening matter in direct proportion to its field gradient.**

** Author's Note

Although a G etheron flux would proceed from the negative to positive pole, this flux is not a direct cause of the force. The G concentration gradient is the perpetrator. The G gradient exerts its force on intervening subatomic particles by stressing their ether concentration patterns. As computer simulations will demonstrate, the particle relieves the applied stress through accelerative movement. Thus, force field action is an example of the operation of Le Chatlier's principle. The electrostatic and gravitational force field equations that result are indistinguishable from those of classical theory.

References

1. Brown, T.T., "How I Control Gravity." *Science and Invention Magazine*, August 1929. Reprinted in *Psychic Observer*, Vol. 37, No. 1, pp. 14-18.

2. "Electrogravitics Systems: An Examination of Electrostatic Motion, Dynamic Counterbary and Barycentric Control." Report GRG 013/56. Aviation studies (International) Ltd., Special Weapons Study Unit, London, February 1956.

3. LaViolette, P.A. "An Introduction to Subquantum Kinetics: Part II. An Open Systems Description of Particles and Fields," *International Journal of General Systems, Special Issue on Systems Thinking in Physics*, Vol. 11, 1985, pp. 295-328.

4. LaViolette, P.A. "A Tesla Wave Physics for a Free Energy Universe." Paper presented at the 1990 International Tesla Society Conference, Colorado Springs, Colorado.

5. LaViolette, P.A. "The Planetary-Stellar Mass-Luminosity Relation: Possible Evidence of Energy Nonconservation?" *Physics Essays*, Vol. 5, No. 4, 1992.

6. Bertalanffy, L. von, *General System Theory: Foundations, Development, Applications*. New York: Braziller, 1968.

A Homemade Power Supply for Electrogravitics Experiments

by Paul A. LaViolette

As an aid to serious-minded experimenters, this article describes a high-voltage power supply capable of yielding about 75 kv DC with a 10 ma capability.

The Power of High Voltage

Electrogravitics experiments may often yield marginal results if the researcher does not have access to a high-voltage power supply that is capable of delivering sufficiently high voltage. T. Townsend Brown repeatedly stated that, in tests of his devices, electrogravitic thrust rose exponentially with increasing voltage. Such a trend is seen in the data obtained in 1952 when the Office of Naval Research (ONR) evaluated Brown's flying discs.[1] An analysis of the ONR data indicates that the speed of the discs increased with voltage (V) according to $V^{5.5}$ and that propulsion efficiency increased according to $V^{4.5}$ (as shown in the log-log plot, *Figure 1*). Efficiency is figured according to the formula presented in the ONR report: ($E = d \bullet v / k \bullet V$), where (d) is the distance between electrodes, (v) is the disc velocity, and (k) is the ionic mobility. The voltage values plotted here are average values from a full-wave rectified AC power supply. (Average voltage is 0.64 of peak voltage.)

As seen here, the ONR investigator took the voltage up to a maximum of 47.3 kv DC (average) at which point the discs propelled themselves at a speed of just 2.8 miles per hour. The ONR data indicate that for these tests,

Brown had chosen a variac that would saturate easily so as to prevent his high- voltage power supply from producing more than 47 kv, and thereby insuring unspectacular results in the ONR tests. As a result, the ONR investigator concluded that Brown's device was impractical for aviation. In reaching this conclusion, the investigator apparently failed to appreciate the fact that the trends in the ONR data project enormous speeds and efficiencies at voltages over 100 kv. Such extrapolations are consistent with the report in *Interavia*[2] that Brown's discs would attain speeds of hundreds of miles per hour when energized with hundreds of kilovolts.[2]

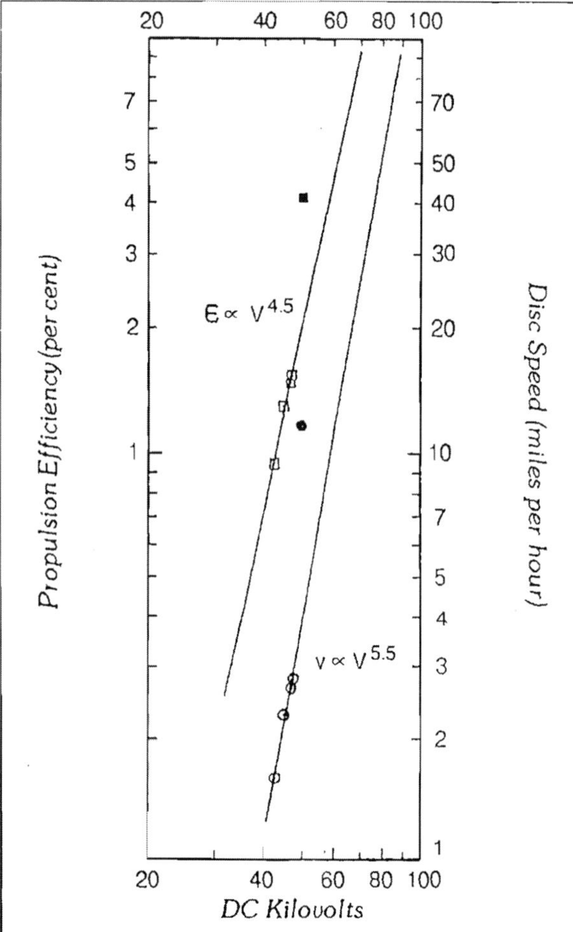

Figure 1. Disc velocity and propulsion efficiency as a func-tion of average DC voltage plotted from ONR test data. Filled points are data for a separate test carried out by Brown.

The filled circle and filled square data points in *Figure 1* plot speed and efficiency for another test in which Brown flew his discs at 12 miles per hour under a charge of 50 kv. Without more information, it is difficult to say why Brown's discs performed better in that test than expected on the basis of the ONR data ($v \propto V^{5.5}$) trend line. One possibility is that this better performance is due to a different disc consruction. In the 1952 ONR tests, Brown used discs made from a circular plexiglass sheet with sheet aluminum paraboloid saucers cupped on either side. On the other hand, the 1956 *Interavia* article, which reports speeds of several hundred mph, describes that the central portion of Brown's discs is made from solid aluminum. Consequently, the better performance might be attributed to a design that utilized a greater inter-electrode mass. If so, this would be in keeping with Brown's claim that his discs are propelled by a gravitic force and not by an ion wind effect.

The ONR test data itself contains evidence favoring a gravitic rather than an ion wind mode of propulsion. Regression analysis of the data indicates that ion current (i) varied as ($i \propto v^{7.8}$) and that the discs' measured thrust varied as ($F \propto v^{6.4\pm1}$). The ONR investigator derived the theretical formula ($F = i \bullet d/k$) to account for thrust in terms of ion wind. Hence the ion wind hypothesis predicts that thrust (F) should vary directly with ion current or according to ($v^{7.8}$). Clearly this exponent is much higher than the observed value. By comparison, if the thrust is due to a gravitic force created by a space charge differential, the thrust would be expected to vary directly with space charge density (ρ), hence according to ($v^{6.8}$). (Space charge ion density varies directly with ion flux (w) as [($\rho = i / (A \bullet w)$)], where (A) is the area cross-section of the ion flux. The ion flux, in turn, varies directly with voltage, ($w = k \bullet v/d$), where (k) is the ionic mobility and (d) is the electrode separation. Substituting for (w) and knowing that ($i \propto v^{7.8}$), we conclude that ($\rho \propto v^{6.8}$.) Since this exponent falls within the range of the observed value of

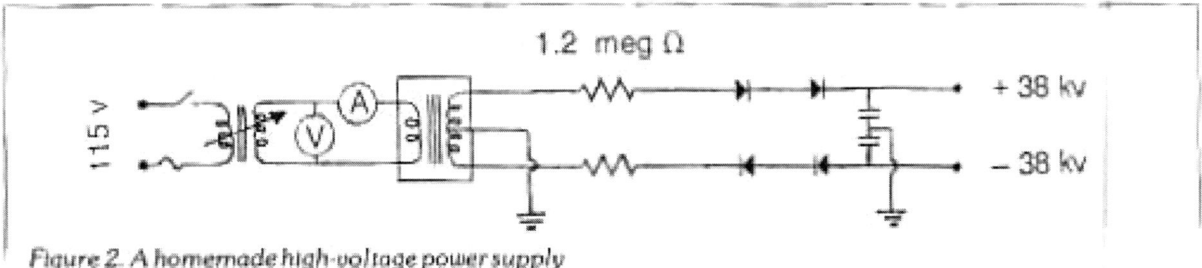

Figure 2. A homemade high-voltage power supply

6.4±1, we must conclude that the gravitic hypothesis constitutes a better explanation.

Building a High-Voltage Power Supply

The principle component of the power supply that I built is the high voltage transformer. The transformer was retrieved from a surplus dental x-ray machine. I used the x-ray head from a pre-World War II William Meyer Co. unit donated by a local x-ray equipment repair shop. The head contained a high-voltage transformer with an x-ray tube connected across its secondary winding, all imersed in an oil bath. Normally, when an x-ray tube in such a unit goes bad, repair people usually replace the entire head and dispose of the old one even though the transformer inside it may still be good. Such equipment is not usually given to individuals due to the lethal consequences of its misuse. So, the repair shop will usually want to know the purpose for which it is intended.

Dental x-ray heads have no provision for external connection to their high-voltage transformer, so it is left to the experimenter to make the proper modification. I converted my transformer as follows. First, I made a wiring diagram of the various transformer connections. I then opened the case, pulled the transformer out of its oil bath, and tested the transformer windings for continuity and to check for any possible short-circuits. I then removed the x-ray tube, and ran high-voltage leads from each side of the transformer's secondary winding through two 7/16" diameter holes drilled in the insulated cover of the transformer. To avoid voltage breakdowns these holes must be positioned so that the terminals are well separated from each other and from the transformer's grounded metal case which connects with the transformer center tap. To further prevent arcing, the high-voltage leads are encased in 15 centimeter-long 10 mm diameter paraffin-filled glass tubes which, in turn, are sealed into their holes with silicone. The ends of these glass tubes must extend sufficiently below the top cover to penetrate several centimeters below the surface of the oil. (See *Figure 3.*)

If possible, try to get an x-ray unit that does not contain PCB type transformer oil, either a unit made very recently or a very old pre-World War II unit. Personnel at the x-ray repair shop can probably advise you on this matter. As a precaution, when working with the transformer, use latex gloves so as to keep the oil from contacting your hands. It is probably a good idea to install a 1/4" diameter fill tube in the transformer top so that additional oil can be added if needed. If transformer oil is not available, mineral oil should suffice.

Once modified to provide external leads, the transformer was connected to the other components, as shown in *Figure 2*, to produce smoothed half-wave rectified DC. Each high voltage terminal of the transformer was connected through a 1.2 megohm limiting resistor to two RCA SK 3443 rectifiers hooked in series, each being capable of rectifying 2.2 mA average current (200 mA surge) and of withstanding a reverse voltage of 45 KV (90 kv for both in series). The rectifiers cost about $9.50 each and

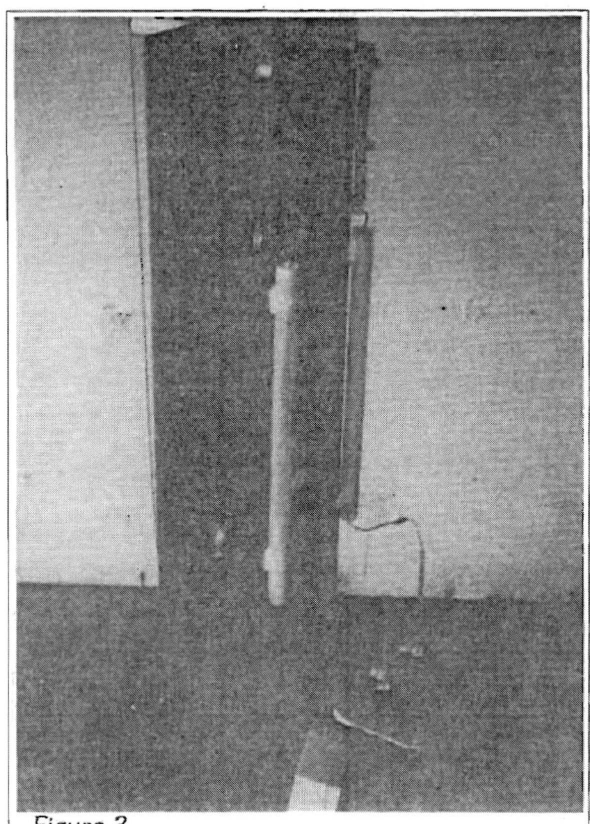

Figure 3

can be ordered from an electronics retailer. Alternatively, Phillips ECG 513 rectifiers may also be used. To build a power supply with full-wave rectification using a bridge circuit, twice as many rectifiers will be needed (a total of eight).

The 1.2 megohm resistors were made by connecting six 200,000 ohm, 1 watt composition resistors in series and insulating them inside a one-inch diameter paraffin-filled glass or plastic tube. Two UHV-10A TDK ceramic capacitors (560 μF, 50 kv DC) were used as filtering capacitors. These were obtained from an electronics wholesaler (All American Semiconductor) at a cost of about $40 each. They retail for much more. All of these high-voltage components were held on stand-off insulators which were mounted to a plywood panel. Power was supplied to the transformer through a cable running to a remote control panel which included a variac, a switch, a fuse, and meters to measure AC volts and current.

Warning

The experimenter should be aware that such high voltages can be extremely dangerous and that direct contact with the transformer's output could result in death. So, it is advised that all high-voltage components be enclosed in a grounded enclosure to prevent accidental contact when the unit is operating. When operated at 115 volts, the transformer yielded 27 kv (rms) on either side of the grounded center tap, allowing the power supply to deliver 76 kv smoothed DC between its output terminals.

References

1) Cady, W.M., "An Investigation Relative to Thomas Townsend Brown," Office of Naval Research, Pasadena, June 1952.

2) "Towards Flight Without Stress or Strain...or Weight," *Interavia Magazine*, Issue 11, No. 12, 1956, p.992.

About the Author

Paul A. LaViolette received his BA in physics from Johns Hopkins University, His MBA from the University of Illinois at Chicago, and his Ph.D. in systems science from Portland State University. Dr. LaViolette is the president of the Starburst Foundation, a research facility investigating astronomy, climatology, psychology, and advanced technology concepts. Dr. LaViolette has special interests in solar energy, archeoastronomy and ancient mythology. He holds several patents and has pioneered the theory of "Subquantum Kinetics."

now in book - by same title. —TV

T. T. Brown Experiment Replicated

Larry Deavenport

Larry Deavenport is a hobbyist in electronics who attended the Texas State Tech Institute. He spent 20 years in maintenance at a copper refining plant.

The first time I recall hearing T. Townsend Brown's name was in a grade school science class during the early 1960s. At that time, NASA was in the process of developing ion and plasma rockets.

Like Nikola Tesla before him, Brown was a pioneer. At an early age, he was to make a remarkable discovery that would lead to the establishment of a field of research called electrogravitics.

While still a student at Denison University in Granville, Ohio, Brown noticed movement from wires of opposite polarity which would come in close proximity to one another. These wires had a tendency to attract while in the presence of very high voltage. Brown reported his observation to his physics professor, Paul Biefield, who encouraged him to do further experiments. The research that followed led to the Biefield-Brown effect and six U.S. patents for Brown.

In one of his experiments, Brown mounted condenser plates on a rotor. When voltage was applied, the rotor turned in a set direction. If the polarity of the voltage was reversed, the rotor would spin in the opposite direction.

The experiment which resulted in U.S. Patent #2,949,550, is perhaps Brown's most well-known. A large disk-shaped negative plate follows a positively charged wire (electrode) separated by a dielectric insulator. When two of these disks are suspended from a rotor mechanism and high voltage is applied, movement follows which is in the direction of the positive electrode, as illustrated in Figure 1.

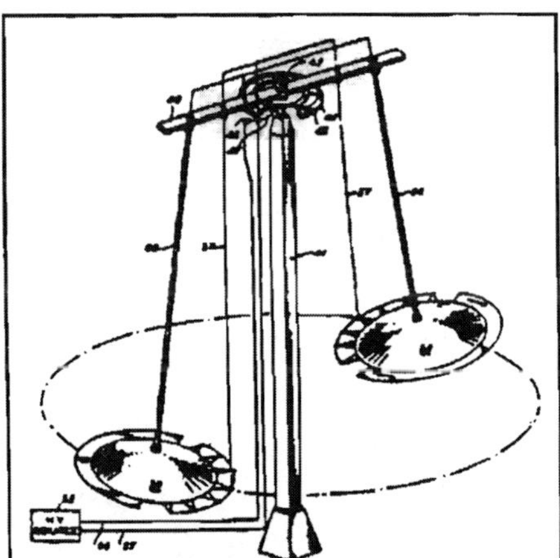

Fig. 1 T.T. Brown's "electrokinetic apparatus." Figure 6 from U.S. Patent #2,949,550, granted Aug. 16, 1960.

To order any of the items mentioned in the article, contact Bob Ianinni at *Information Unlimited*, P.O. Box 716, Amherst, NH 03031-0716. Tel 603-673-4730.

The *IU* catalog is $1. The plan is part GRA1, and sells for $15. The kit is part GRA1K, and sells for $99.50. The 12V, 3amp power supply is part GRA10, and is made and priced to order.

The replication

During the summer of 1994, I decided to replicate Brown's rotor experiment. I machined two 2⁵/₈-inch by ¹/₈-inch thick aluminum disks, with a tapered edge.

Using a 16-inch by 2-inch wide piece of balsa wood, I built a crude, but workable, rotor to suspend the disks. Small pieces of #36 steel wire, separated by insulators, were attached to the rotor as condensers. A strip of ³/₈-inch copper was formed into a ring and glued to a 1¹/₂-inch by 18-inch tall PVC support for brush contact. The rotor was then suspended on top of the support with a glass door roller used for the bearing. The bearing was later replaced with a needle spindle because the original caused drag problems by creating too much surface area, which prevented the rotor from turning.

The power supply was a 0-15 volt input oscillator with a

25kV, 800 microamp output at 25kHz pulsed DC. The circuit diagram is illustrated in Figure 2.

When I applied the voltage, the disks behaved as if they wanted to turn, but they did not. I made several adjustments between the positive and negative electrodes of the disk, but still there was no rotation.

At the same time, I was experimenting with other power supplies. In one of these other trials, a 7000VAC, 5 milliamp transformer with six 10,000V capacitors and six 20,000V, 2 watt diodes were used. The six capacitors and six diodes were soldered together to form a Cockroft-Walton generator. This was set with spark gaps between them, and run in parallel with the original power supply. When the power supplies were turned on, the higher current transformer with the Cockroft-Walton DC ladder began to amplify the action of the lesser powered 25kV, 25Hz pulsed DC supply. The aluminum disks began to resonate with every aluminum box in the room. The disks behaved as if they wanted to move up and down, but the rotor did not turn.

It was several months before I tried any more experiments. Early this year, I was discussing the Brown experiment and the problems I was having with a friend. I showed him the rotor and mentioned the problem of having too much bearing surface drag from too large a bearing. We agreed that the bearing should be smaller, and decided to try a needle or small spindle. The changes were made, and on the first weekend in February, I started experimenting again. The new setup is shown in figures 3 and 4.

This time, the small rotor turned more freely; however, it needed to be balanced. Bearing surface drag and rotor balance will greatly affect the outcome of the experiment. Also very important is the size of the rotor bearing and how well the unit is balanced in conjunction with the size of the power supply.

The rotor began to turn slowly at first, at about 15 rpm. Several days later, I repeated some experiments with the rotor, and found that a ground connection was loose. When this was corrected, the rotor speed increased to about 35 rpm.

During this time, I contacted Dr. Paul LaViolette who suggested making the center of the disks more blunt instead of tapered. On my new 4-inch disks, shown in Figure 5, I took this into consideration.

One of the later experiments with the disks used a much larger power supply—0-50 volts DC, 0-15 amp input. The voltage was increased to

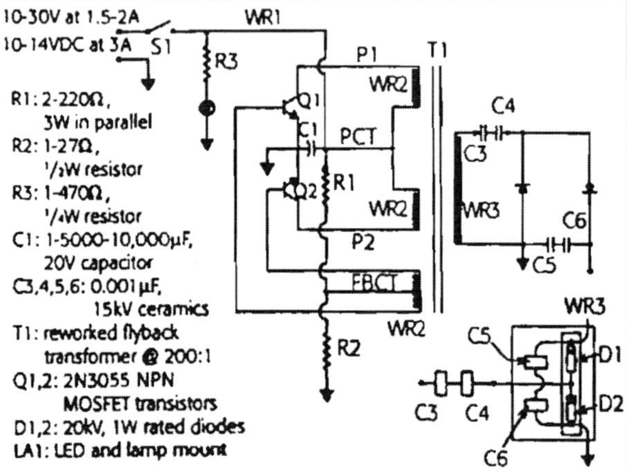

10-30V at 1.5-2A
10-14VDC at 3A S1
R1: 2-220Ω, 3W in parallel
R2: 1-27Ω, 1/2W resistor
R3: 1-470Ω, 1/4W resistor
C1: 1-5000-10,000μF, 20V capacitor
C3,4,5,6: 0.001μF, 15kV ceramics
T1: reworked flyback transformer @ 200:1
Q1,2: 2N3055 NPN MOSFET transistors
D1,2: 20kV, 1W rated diodes
LA1: LED and lamp mount

Fig. 2 Circuit diagram for rotor.

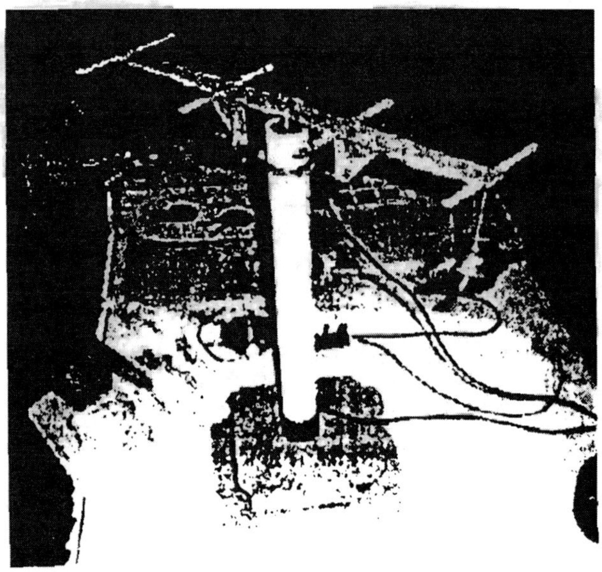

Fig. 3 The author's "maypole" rotor device.

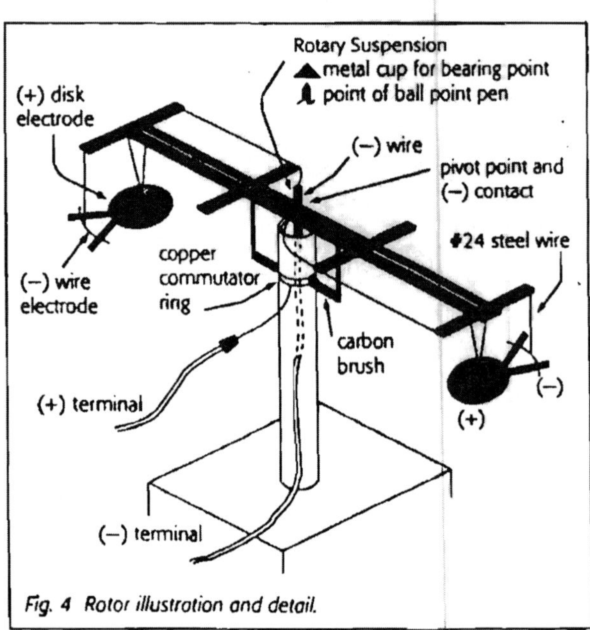

Fig. 4 Rotor illustration and detail.

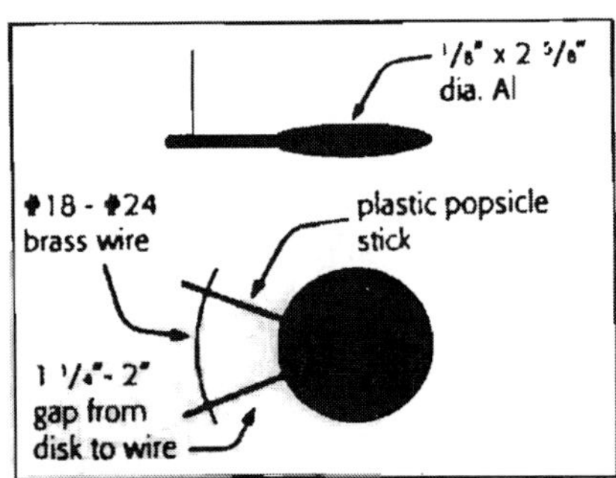

Fig. 4
Enlarged view of the disk electrode.

various levels and special notice was taken as to what happened on the high voltage end. The original power supply or oscillator was rated for 6-15 volts with a 25 kHz DC pulsed output. It was originally purchased about 10 years ago for the Hagen device (U.S. Patent #3,120,363). In this experiment, a platform about 12 inches in diameter is negatively charged, while a smaller circular platform containing very small wires is positively charged. The slow moving mass of air and charged ions are supposed to generate lift. Regardless, this power supply worked well on Brown's experiment.

I also contacted Bob Ianinni, the owner of Information Unlimited and the manufacturer of this device. He first thought that due to the age of the power supply, a pulsed rippling effect might be generated from the capacitor. Since I do not own an oscilloscope, and have only a linear frequency counter built into my meter, there is no proof that this may have happened. However, after conducting tests using other capacitors (a 63 volt, 100,000 µF and a 63 volt, 15,000 µF) in place of the original 20 volt, 10,000 µF capacitor, I ruled out the rippling effect. If there had been a pulsed ripple produced from the capacitor, it would not have been present in a larger capacitor.

In the higher voltage experiments, the speed of the rotor increased significantly, from 35 rpm to about 60 rpm. This is after the power supply had reached 24 volts, according to the VOM reading. I checked with an analog meter for verification of input. This would have doubled the output of the high voltage to 50,000 volts and the current to about 1.5 milliamps. The rotor turned at a maximum speed of 60 rpm.

Several factors were responsible for the first breakdown between electrodes on the disks. High voltage power transistors had reached their maximum wattage, causing overheating, and the needle bearing on the rotor prevented the rotor from reaching high speeds.

Another experiment involved adding the Crockroft-Walton voltage ladder to the high voltage terminal of the power supply. This stepped up the voltage, but dropped the microamp rating down. The rotor turned at about half the speed it normally turned, at 25kV. This indicates that voltage and current rise must remain proportional to maintain the effect.

An analysis of Brown's experiments

1 Voltage and amperage must be raised proportionally to maintain the electrogravitic effect.
2 As the voltage and current are raised, dielectric strength (K) or spacing between positive and negative electrodes must be increased.
3 Power or frequency of charging time must be increased proportionally as the size of the area is increased.
4 Changing the size of either electrode will affect the efficiency of ionic flow, if the dielectric and power source remain constant. The ratio in the size of the positive to the negative plate is directly proportional to the amount of voltage and current applied through a dielectric medium. With a smaller, high voltage low microamp power supply, the electrode on the positive side may have to be larger. On a higher voltage milliamps system, the positive electrode may need to be smaller.
5 Geometric configurations of shape may enhance or diminish the efficiency of the electromagnetic effect.
6 The electromagnetic effect increases in proportion to voltage, amperage, dielectric strength (the K factor) and area, when all of these increase together. It is lost when an arcover or breakdown occurs.
7 Some writers have, in the past, insinuated that electronic levitation is a product of very high frequencies and high voltages. Although a significant power increase may be present in this state, my experiments with the Biefield-Brown effect show that it is determined by the conditions thus described. Certainly, frequency, time lag and phase angle, along with all the formulas dealing with resonance would affect the efficiency of an AC system if it were used, but frequency and high voltage by themselves do not produce this effect.

In conclusion, I believe that the electrogravitic force is ever present, be it at one volt, one amp, one ohm, or 50,000 volts and 1 milliamp with a high dielectric strength. It is simply more noticeable at the high voltage electrostatic state, which is the opposite of the electromagnetic state.

Contact the author, Larry Deavenport, at 1823 Lawson Lane, Amarillo, TX 79106.

Thomas Townsend Brown was a physicist, inventor, and researcher whose work is largely unknown and hard to find because it does not fit easily into mainstream science. Brown invented the "gravitor," a device that demonstrated what was called the Biefeld-Brown Effect. This effect is defined as "the observed tendency of a highly charged electrical condenser to exhibit motion toward its positive pole."

T.T. Brown was intensely interested in demonstrating electrogravitational phenomena. His lab notes testify to the constant and fundamental nature of his quest.

THE ELECTROKINETIC WORKS OF T.T. BROWN

by Steve Hall

INTRODUCTION

The purpose of this article is to present information on the little-known research efforts of T.T. Brown, particularly during the years 1957 through 1960. *Electric Spacecraft Journal* has reviewed a piecemeal video excerpted from a lab film of actual experiments performed by Brown, Agnew Bahnson, J. Frank King, and other associates of the Bahnson Labs in Winston-Salem, North Carolina.

Brown's main goal was to develop an antigravity propulsion system that would be spaceworthy. This lifelong focus led Brown across the United States and Europe in hopes of finding support and funding for his work.

Detailed information concerning Brown's research is difficult to find. If anyone has details from the Brown experiments, the *Electric Spacecraft Journal* would appreciate hearing from you.

HISTORICAL

T.T. Brown was born in 1905 to a prominent family in Zanesville, Ohio.

Sometime between 1913 and 1922 Brown obtained a Coolidge X-ray tube, with the thought that a key to space flight may be contained in this device.

His first experiments were set up to test for any type of forces that might be exerted by the x-rays to propel an object. He noticed that each time he turned on the tube it seemed to try to move on its own accord.

Brown explored this effect until he devised what he called a "gravitor." When connected to a 100 kv/dc power source, it lost 1% of its weight. He developed this instrument even before graduating from high school.

In 1922 Brown entered California Institute of Technology (Caltech) in Pasadena for one year, then transferred to Kenyon College in Gambier, Ohio. In 1924 he transferred again, to Denison University at Granville, Ohio, where he came under the guidance of Dr. Paul A. Biefeld, who was professor of physics and astronomy. (Dr. Biefeld was one of the eight classmates of Dr. Einstein in Switzerland.)

Brown and Dr. Biefeld experimented with charged capacitors and developed what is called the Biefeld-Brown Effect. This effect is observed when a high voltage capacitor is charged and exhibits a motion toward the positive plate or pole. Over the next four years Brown worked as a staff member at the Swazey Observatory with Dr. Biefeld. Also at this time he married and started his family.

In 1929 Brown wrote his first serious paper, "How I Control Gravitation," published in *Science and Invention*. Still, no one took his work seriously.

In 1930 he took a position with the Naval Research Lab (NRL) in Washington, DC. There he worked in areas of radiation, spectroscopy, and field physics. During this time he took part in the U.S. Navy's International Gravity Expedition. He also participated in the Johnson-Smithsonian Deep Sea Expedition.

By the year 1933, government funding cuts forced Brown to seek work elsewhere. He joined the Naval Reserve at this time and worked at any science-related jobs that became available.

In 1939 Brown and his family moved to Baltimore, Maryland, where he started working for the Glenn L. Martin Company (Martin Aerospace) as a materials engineer. This job was short-lived because the Navy called him to active duty. He became the officer in charge of magnetic and acoustic minesweeping research and development under the Bureau of Ships. It is speculated that at this time Brown became associated with the "Philadelphia Experiment."

Shortly after Pearl Harbor he was transferred to Norfolk as a lieutenant commander in charge of the Navy's Atlantic Fleet Radar School. Again it was reported that he was involved with the Philadelphia Experiment, working with electromagnetic fields to achieve radar invisibility.

Tragedy came in December of 1943 when Brown suffered a nervous breakdown. After being sent home to convalesce, he was soon retired from the Navy, upon a naval physicians' review team recommendation.

By the spring of 1944 Brown was employed by the Lockheed-Vega Aircraft Corporation in California as a radar consultant. He then left Lockheed and moved to Hawaii, where he worked temporarily as a consultant at the Pearl Harbor Navy Yard.

During all this time Brown had continued to improve on his gravitor, and by 1951 had produced many working models. Brown had already envisioned the use of the Biefeld-Brown effect to propel air- and spacecraft and to act as a gravity field manipulator. The movement of ionized air molecules by high voltage fields was already known to produce propulsion. In fact, Brown had pointed this out in his patent #2949550. What he did not point out were his thoughts of direct gravity field interactions.

In 1952 Brown moved to Cleveland, Ohio and started his own project called "Winterhaven," which dealt with electrogravitics.

In 1953 he demonstrated the movement of disk-shaped electrokinetic transducers. They were two feet in diameter and reached a speed of 17 feet per second when a potential of 50 kv/dc was applied through a flexible lead, allowing the craft to circle a 20 foot-diameter path.

Brown continued developing better output craft, but due to a lack of interest in the U.S. he left for Europe in the hope of finding the support and funding needed to continue his research. In France he conducted research at the La Société Nationale de Construction Aéronautique Sud Ouest (SNCASO). Here he was able to increase the potentials to at least 200 kv/dc.

By 1956 Brown was again disappointed when the company SNCASO merged with Sud Est and funds were cut, forcing him to return to the U.S. Upon his return he founded the organization NICAP (National Investigative Committee for Aerial Phenomena) to study the Unidentified Flying Objects which seemed to him to exhibit many of the qualities of flight he was trying to attain.

By 1957 Brown was working as chief research and development consultant for the Whitehall-Rand Project for the Bahnson Company in Winston-Salem, North Carolina. The project was under the direction of Agnew Bahnson, president of the company. The project was set up to study antigravity. It is this period in Brown's life where the film mentioned earlier should shed some light on his experiments.

In 1959 Brown started working at the General Electric Space Center in King of Prussia, Pennsylvania. He worked there until at least June of 1960, according to the film footage. Brown also set up his own company called Rand International. The association between Bahnson and Brown is not known after 1960, but again tragedy struck when Agnew Bahnson was killed in a 1964 plane crash, and the project was suspended by the company.

Brown went into semi-retirement sometime in the 1960's and returned to Avalon, California to live.

As late as 1983 Brown was still involved in research efforts with a firm in Los Angeles. There he concentrated on a new form of radiation from space which was a carryover from his work with the Naval Research Lab in 1930. Cal State and the University of Hawaii were also involved. This new radiation held some promise in communications underwater as well as through large objects like a planet.

The life of Thomas Townsend Brown came to a close in October of 1985, at Avalon. A great deal of research information remains unknown.

T.T. Brown's Patent

One of Brown's patents, #3,018,394, which was filed on July 3, 1952 and patented on January 23, 1962, says in part:

> This invention [Electrokinetic Transducer] utilizes heretofore unknown electrokinetic phenomenon which I have discovered, namely that when pairs of electrodes of appropriate form are held in a certain fixed spacial relationship to each other and immersed in a dielectric medium and then oppositely charged to an appropriate degree, a force is produced tending to move the surrounding dielectric with respect to the pair of electrodes.
>
> I have also discovered that if the dielectric medium is moved relative to the pairs of electrodes by an external mechanical force, a variation in the potential of the electrodes results which corresponds to the variation in the applied mechanical force.
>
> Accordingly, it is an object of this invention to provide a method and apparatus for converting the energy of an electrical potential directly into a mechanical force suitable for causing relative motion between a structure and the surrounding medium.
>
> It is another object of this invention to generate a varying electric potential by the motion of a surrounding dielectric medium with respect to a pair of electrodes supported in the medium.

The patent also explains other aspects of the invention, as well as its construction. As in his other patents, positive-negative pairs of electrodes are connected together in groups. If these electrodes are suspended so as to allow their movement, the movement is in the positive pole direction when high voltages are applied. If the pair(s) of electrodes are securely mounted so as not to allow movement, then the medium is forced to move with respect to the positive electrode. In the patent, Brown states that this is analogous to a pump or a fan.

FILM REVIEW

Electric Spacecraft Journal has reviewed approximately 45 minutes of the Bahnson/Brown laboratory experiments on film clips (videotape copy). There was no sound or lab notes accompaniment. Lab notes do exist elsewhere. We have taken a few selections from the film and present them here as examples for others to comment upon.

Photo #1 is the first electrokinetic transducer to be discussed. This picture extracted from the video is provided for reference. The most noticeable feature is the large umbrella-shaped electrode at the top. There are approximately 10 to 12 metallic ribs supporting the canvas-like covering. The material used is not known, but appears to be either cloth or a latex of some sort.

At the end of each rib appears to be a disk-shaped insulator. These insulators seem to be 2 or 3 inches in diameter and spaced about 6 inches apart. Each insulator has a hole in the center through which a 1/2 to 3/4 inch metallic band passes. This band encircles the entire perimeter and, based on the color, is most likely copper.

There is also a plastic film, like cellophane, covering the area from the bottom of the canvas to the bottom of the insulators.

There are four cross-bars connecting radially from the center to the top electrode perimeter. It is not clear, but it seems that these four arms or rods are connected to the four rods that connect the top and bottom electrodes. The assumption would be that the top electrode is the positive due to its size, and thus the bottom one would be the negative electrode. The composition of the connecting rods is not known. At this time it is not known whether these serve as insulators or in some other capacity.

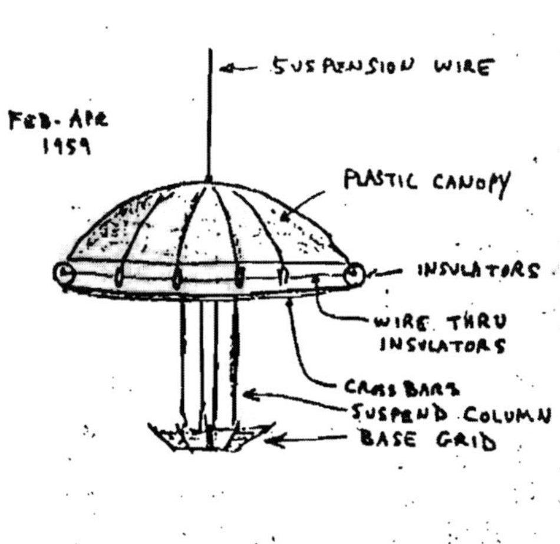

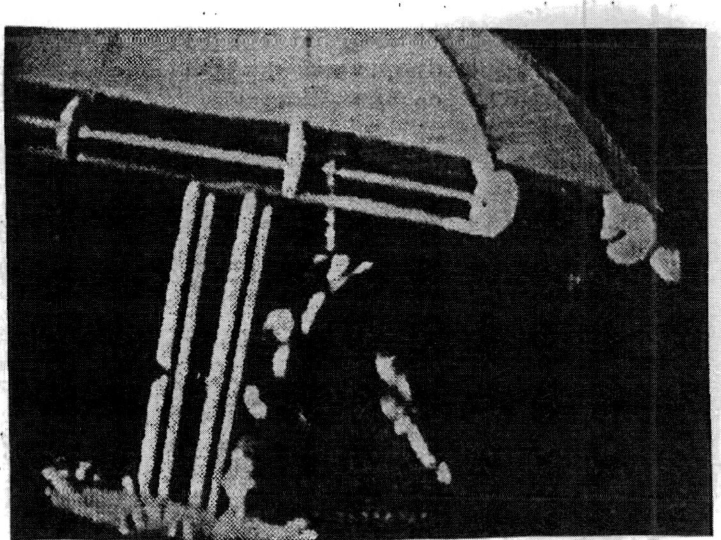

Photo #1

The bottom electrode, in contrast to the top one, is many times smaller and appears to be all metallic in composition. It also has about 8 ribs which seem to be of the same material. They do not have any insulators attached nor can any plastic film be discerned around them.

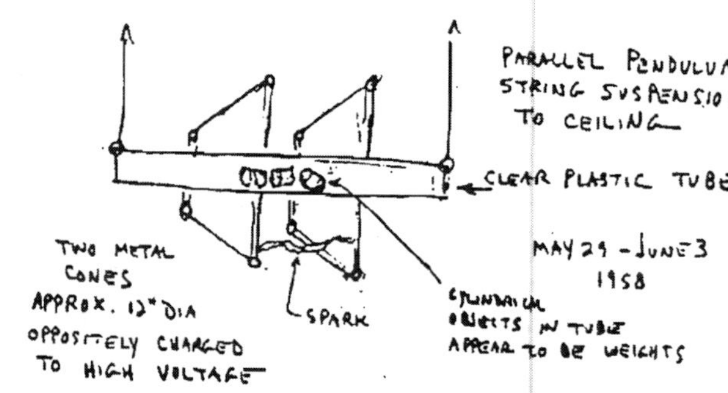

As with most of the experiments observed in the video, there are flexible leads connected to one or both of the electrodes and sometimes an additional weight is also attached to the lead. Details about the power source are unknown. From what is known and observed, it appears that a very high voltage DC was the main power utilized.

Photo #2

This apparatus was shown at first to be hovering in place, bobbing slightly. Then, as if the voltage is increased, the object quickly moved up and the video moves to a calendar where someone circles the date February 28, 1959.

Photo #2 is a different type of experiment. The electrodes are similar in size and shape and are mounted to the floor. These open-ended, conic-shaped electrodes are used extensively, both in the atmosphere and in a vacuum chamber in other experiments.

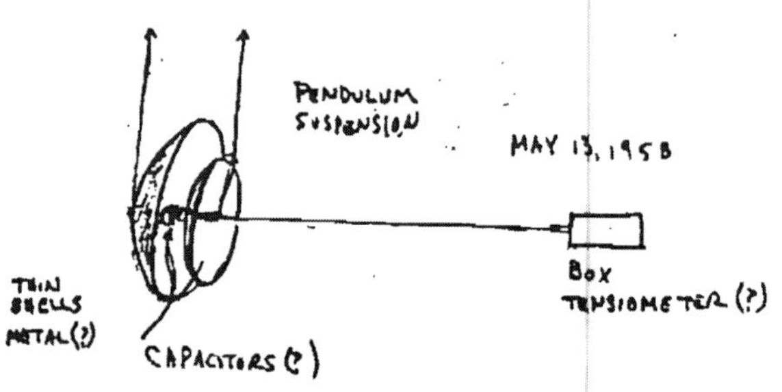

Suspended by cords from the ceiling is a horizontal glass or plastic tube positioned to swing through the openings in the conic electrodes. Inside this tube are three squat cylinders which appear to be connected but do not extend outside the tubes. Whether these three objects are dielectrics, series-connected high voltage capacitors, or simple weights is not known.

High voltage power is applied to the conic electrodes and the tube swings back and forth. The motion is like a parallel pendulum in the high voltage field.

Photo #3

Brown may have been testing for variations in a pendulum swing that might show an interaction of the electric and gravity fields.

Photo #3 is included to show three squat cylinders mounted in series. These cylinders are in all likelihood high voltage capacitors between two dome-shaped electrodes. (Light-weight dome-shaped electrodes are used in many experiments, including those used in the vacuum chamber.) The whole apparatus is suspended from the ceiling with cords similar to those used in Photo #2. A horizontal cord is connected to what may be the negative electrode or a meter of some sort.

When power is applied, the suspended shells move toward the front or positive electrode. With this happening, it is possible to conclude that a self-propulsion device was successfully tested. It should be pointed out that Photo #1 also exhibits this effect. It should also be added that this apparatus could have been testing the pendulum variations as seen in Photo #2.

Photo #4 shows a disk that seems to be similar to some UFO objects. This could be deceiving though. The apparatus is shown only for a few seconds and does not exhibit any motion. It is simply lying on the table with the feet of a person shown in the background.

On top is mounted a small diameter glass or plastic disk. On the bottom is a larger disk of similar material. Sandwiched in between is a saucer-shaped metallic disk. This is similar to a flying saucer, but is, in fact, another electrode/dielectric configuration.

There are many unknown variables involved which make drawing any conclusions difficult at this time. Lab note data exists, but *Electric Spacecraft Journal* has not seen these to date. What tests were being conducted, and what results were expected are unknown, with the exception that T.T. Brown was determined to produce an electrokinetic transducer that could propel itself. If anyone has further information on the figures shown, other experiments or data, please contact us at the *Electric Spacecraft Journal*.

Photo #4

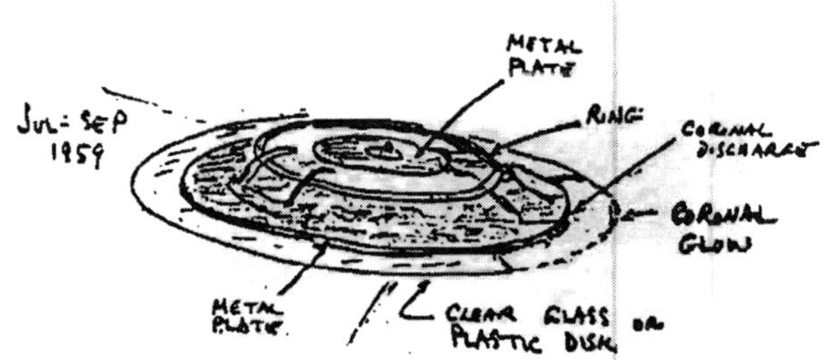

BIBLIOGRAPHY

The Philadelphia Experiment: Project Invisibility by William L. Moore and Charles Berlitz, Ballantine Books, New York, 1979.

Ether-Technology: A rational approach to gravity-control by Rho Sigma, CSA Printing and Bindery, Lakemont, Georgia 30552, 1977.

INTRODUCTION TO PROJECT MONTGOLFIER REPORTS
by Jacques M. Cornillon

http://projetmontgolfier.info/INTRODUCTION.html

I came into the life of Thomas Townsend Brown, in the mid-1950's. I was the U.S Technical Representative for a French aeronautic company, Sociéte National de Construction Aeronautiques du Sud-Ouest, (S.N.C.A.S.O). I was in the library of the Institute of Aeronautical Science of Los Angeles, studying a magazine when my attention was drawn to the word "Wow" written in the margin, with exclamation points. I read the article "The Flying Saucer: A Simplified Explanation of the Application of the Biefeld-Brown Effect to the Solution of the Problem of Space Navigation" by Mason Rose, Ph.D., that said a Townsend Brown had found a new force that seemed to indicate a new form of propulsion for aviation. In this article he outlined the Biefeld-Brown Effect, electrogravitation, and its application to space navigation. There was considerable interest throughout the world on this subject at that time, with many applications for patents. An address was provided at the end of the article, so I decided to contact the author, Dr. Mason Rose. I did, and we met. Subsequently he introduced me to a 'Dr. Shank'. We spoke at length about Brown's project and its history and of my company and our potential interest in Brown's project. At the conclusion they recommended that I contact Brown, which I did on my return to the East Coast.

I met with Townsend Brown in Washington on April 7, 1955. I first asked him if he was free to work with our French company, that he was not already engaged with an American company or with the American government. He told me that he would get back to me in a few days. Several days later, he called me to tell me that he was able to work with us. We then made a contract with him to come spend some time with us at our facility outside of Paris to make the experiments that he had outlined for us. He had given us plans of some 'flying saucers' that he wanted us to build in our workshops, and to procure a machine to produce the needed energy source, (see T. Townsend Brown Proposal and T. Townsend Brown Blueprints).

Our company undertook the project in secret with a limited team, at a building in our Courbevoie facility called 'B12'. We baptized this project "Projet Montgolfier", in honor of the French brothers who were the first men to experience free flight. Our first goal was to make a complete scientific study of the Biefeld–Brown effect.

Dr. Brown came to France twice in the period from 1955 to 1956. Many tests were made. The first round of testing was done in the air. We found that when we 'flew' the saucers around the tether, this was simply a pure demonstration of 'electric wind'. We found there was a small force that sometimes was added to the expected force, but it was much too small to measure in the air where there is ionization.

So it was decided that a new round of tests needed to be made in vacuum to eliminate the effect of ionization in the air. We created a vacuum in a small bell jar with a modified miniaturized saucer and tether design from Dr. Brown and performed a series of new tests. While we did achieve results indicating a separate force not attributable to ionization, the experiments in the small bell jar were affected by the outside atmosphere on the bell jar

wall and rendered these experiments scientifically inconclusive.

It was decided that the next step was to make tests in a big vacuum chamber. Dr. Brown again sent us designs for the construction of a large vacuum chamber and test apparatus.

As this phase of the project was undertaken my company was merged into another company. During this turbulent period of the merger we were able, with difficulty, to continue and complete the construction of the large vacuum chamber, though moved to a less hospitable location. The president of my company, now the president of the new merged company, Sud-Aviation, decided not to continue the experiments but to pass them along to another company S.N.E.C.M.A. (Société Nationale d'Étude et de Construction de Moteurs d'Aviation) that was more specialized in this type of research.

The team made some hasty tests before having the project shut down for delivery of the vacuum chamber to the new company. The Final Report for the Projet Mongolfier, April 15, 1959, outlined these five tests confirming, as in the prior tests, that there was a definable force. At this point our team was scattered, the project shut down and we were unable to make the further tests to further refine and quantify the results.

It was at this point that I lost contact with the project, and despite later efforts was never able to find out what, if anything, happened to it.

Ed. Note: Jacques Cornillon passed away on May 30, 2008, at the age of 99, not long after he concluded these notes.

Hyperlinks in document:

Thomas Townsend Brown → http://en.wikipedia.org/wiki/Thomas_Townsend_Brown

Sociéte National de Construction Aeronautiques du Sud-Ouest, (S.N.C.A.S.O → http://en.wikipedia.org/wiki/SNCASO

Sud-Aviation → http://en.wikipedia.org/wiki/Sud_Aviation

"The Flying Saucer: A Simplified Explanation of the Application of the Biefeld-Brown Effect to the Solution of the Problem of Space Navigation" by Mason Rose, Ph.D., →

http://projetmontgolfier.info/uploads/The_Flying_Saucer-Mason_Rose.pdf

T. Townsend Brown Proposal → http://projetmontgolfier.info/uploads/TTB_Proposal.pdf

T. Townsend Brown Blueprints →
http://projetmontgolfier.info/uploads/TTB_Proposal_Blueprints_1-3.pdf

T. T. Brown's 1955-1956 Paris Experiments Revealed

Posted on May 13, 2012 by P. LaViolette

Townsend Brown flying his discs at the S.N.C.A.S.O. facility outside of Paris. (photo courtesy of J. Cornillon)

In 1955 and 1956 Townsend Brown made two trips to Paris where he conducted tests of his electrokinetic apparatus and electrogravitic vacuum chamber tests in collaboration with the French aeronautical company Sociétė National de Construction Aeronautiques du Sud Ouest (S.N.C.A.S.O.) . He was invited there by Jacques Cornillon, the company's U.S. technical representative. The project was named Project Montgolfier in honor of the two French brother inventors who performed early aircraft flights. The project continued for several years until the company changed ownership resulting in a final report which was written up in 1959.

Details of the Project Montgolfier experiments remained a closely guarded secret for many years until Jacques Cornillon courageously decided to make them public prior to his death in July 2008. Brown's proposal, the project's top secret final report, and an assortment of revealing diagrams and photos are posted on the Cornillon website at:

Project Montgolfier: http://projetmontgolfier.info/

and are available for free download. Brown's proposal is in English, whereas the secret Montgolfier Project final report is in French. An English translation of this 100 page final report is available from the Peeteelab website for a fee of $5.95 Canadian. (Note there is a translation error in the statement of the electrokinetic disc wire size, which is 100 times smaller than quoted. Refer to the original French document for the proper size.)

The flying disc carousel experiment that the Montgolfier Project conducted in 1955 used 2-1/2 foot diameter discs (75 cm dia.) hung from 4 meter tethers suspended from the ends of a 3 meter arm. Based on the description given, this seems to have been almost the same flying disc test that Brown gave to the Navy at Pearl Harbor a year or two earlier.

Based on the angle of the disc suspension cable seen in the photo on the right below, one may estimate that the disc was traveling at a speed of ~8.7 meters per second, or about 20 mph. It would have completed one revolution of its 18 meter course in 2 seconds.

Brown had finished his collaboration with S.N.C.A.S.O. in 1956. From a letter that Mr. Cornillon later wrote to a colleague, we learn that in October 1957 Brown was in the process of test flying 10 foot diameter discs energized at

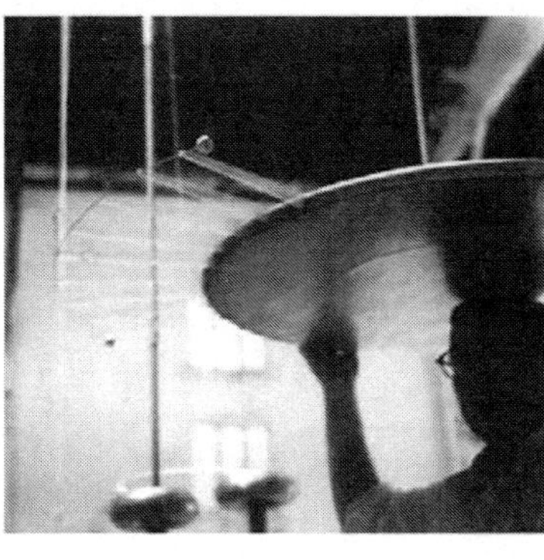

Left: Brown holding a flying disc tested in Project Montgolfier. Right: Close-up of disc showing outboard leading-edge wire. (photos courtesy of J. Cornillon)

a voltage of 300 kV! Here we see that Brown had followed the plan he had first set out in his 1952 Project Winterhaven proposal which was to eventually test fly a ten foot diameter disc powered by 500 kV (70% more voltage than he used in his 1957 test flight). Hence we see that by this early date Brown had progressed beyond the toy model stage to flying small scale aircraft. To reach this stage he must have been receiving substantial funding from either the military or from a major corporation. More about Project Winterhaven and Brown's research may be found in the book *Secrets of Antigravity Propulsion*.

Left: Carrousel test rig. Right: Disc in flight. (photos courtesy of J. Cornillon)

In addition the Project Montgolfier team constructed a very large vacuum chamber for performing vacuum tests of smaller discs at a pressure of 5 X 10^{-5} mm Hg; see below.

In reading the section describing the vacuum chamber results, we learn that when the discs are operated at atmospheric pressure they move in the direction of the leading edge wire regardless of outboard wire polarity. This indicates that in normal atmospheric conditions the discs are propelled forward primarily by unbalanced electrostatic forces due to the prevailing nonlinear field configuration (which causes thrust in the direction of the low field intensity ion cloud regardless of the ion polarity). On the other hand, the report says that under high vacuum conditions the discs always moved in the direction of the positive pole, regardless of the polarity on the outboard wire. This indicates that in the absence of the unbalanced forces exerted by ion clouds, the discs moved mainly on the basis of the electrogravitic field effect, always toward the positive (negative G) direction.

These vacuum chamber experiments were a decisive milestone in that they demonstrated beyond a doubt that electrogravitic propulsion was a real physical phenomenon. The report concludes saying: "It seems perfectly reasonable to conclude that a concentrated force of some kind accumulates within the presence of a strong dielectric." (i.e., presumably in the presence of a high-K dielectric.)

This entry was posted in Townsend Brown. Bookmark the permalink.

Left: Vacuum chamber vessel (1.4 m diameter) for conducting electrogravitic tests. Right: Vessel opened to show test rotor rig within. (photos courtesy of J. Cornillon)

Starburst Forum: Electrogravitics and Field Propulsion
Proudly powered by WordPress.

"Zero-Point" Documentary on Electrogravitics Fluxliner

IRI Press Release March 23, 2014, *Future Energy eNews*
http://www.youtube.com/watch?v=ZQfofTXIvb8

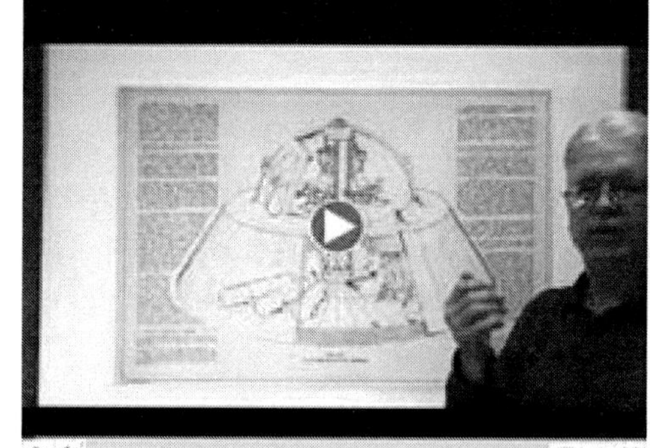

Bolide Motion Pictures in association with the Centre for Jim Gladman Studies released a documentary about a military vehicle illustrator, quantum physics and possibly reverse-engineered alien propulsion technology, currently unfinished. The first five minutes are an introduction and overview of the Disclosure Project from 2000 and the rest is a full length interview and summary of the pulsed electrokinetics hovercraft that was on display at an Air Force Base in 1988. It appears from the performance capabilities of each craft that "inertial shielding" is also involved in the design of the hull of the craft to account for its reported flight cabilities. Mark contributed to a second book in the series, ***Electrogravitics II: Validating Reports on a New Propulsion Methodology*** by Thomas Valone, on Amazon.com and published by IRI. It reviews the craft seen in the last chapter of the book, with Mark's contributed article on the craft, which was on display at Norton AFB in 1988, which is powered by pulsed electrokinetics that is explained by Prof. Jefimenko's electrokinetic equation.

DATE-EVENT CHART FOR T.T. BROWN

1905	March 18, TT Brown born in Zanesville, Ohio.
1921	TTB discovers motion of the Coolidge X-ray tube caused by high voltage.
1922	Attended Cal-Tech; left school and joined the Navy.
1923	Attended Kenyon College, Ohio.
1924	Attended Dennison University, Gambier, Ohio where he met Prof. Biefeld (Ph.D., Zurich, Switzerland, 1900).
1926	Joined Swazey Observatory in Ohio, directed by Prof. Paul Alfred Biefeld.
1928	Maximum thrust measured on "Gravitor," 1% of total weight.
1929	Wrote article for *Science and Inventions*, "How I Control Gravitation."
1930	Left Swazey Observatory, signed on to Naval Research Lab, Washington, DC.
1932	U.S. Navy Department, staff physicist, International Gravity Expedition to West Indies.
1933	Staff physicist, Johnson-Smithsonian Deep Sea Expedition; joined Naval Reserve; soil engineer for Federal Emergency Relief Administration; joined FCC as an administrator.
1939	Lieutenant in Navy Reserve, moved to Maryland as a materials engineer for Glen L. Martin (Aircraft) Co., called into Navy Bureau of Ships as an officer in magnetic and acoustic mine research and development.
1940	May have joined the "Philadelphia Experiment."
1942	Lt. Commander, Commanding Officer of U.S. Navy Radar School at Norfolk, VA.
1943	Collapsed from exhaustion, retired from Navy, 6 months recuperation. Gravitor not recognized.
1944	Position with Lockheed-Vega, radar consultant.
1945	Went to Hawaii, continued research: (Presented ideas to Admiral Arthur W. Radford, Commander in Chief of U.S. Pacific Fleet, who later became Joint Chief of Staff/Eisenhower, 1953-57. No apparent interest was shown).
1952	Started Winterhaven Project in Cleveland, Ohio. Set up Townsend Brown Foundation office in Menlo Park, CA and demonstrated "flying saucer-"shaped disks electrically propelled around a "maypole."
1955-56	Presentations in England of his ideas.
1955-57	Leesburg, VA from Oct. 1955 to Feb. 1957; Umatillo, FL, Nov. 1957.
1956	Project work in France with his electrogravitation ideas for SNCASCO. Brown formed UFO study group, NICAP, in Washington, DC.
1957-60	Hired as consultant for Whitehall-Rand Project under auspices of Bahnson Labs (Agnew Bahnson) to do antigravity research and development.
1958	Brown formed Rand International, Ltd.
1958-67	Brown indicated he made no notes in his notebooks.
1959	Consultant with G.E. Space Center at King of Prussia, PA.
1960's	Physicist for Electro Kinetics at Bala Cynwyd, PA. Also semi-retired.
1967	T.T. Brown Notebook: October 23, 1967 in Santa Monica, CA.
1970	T.T. Brown Notebook: May 31, 1970 at Stanford University Hospital.
1973	T.T. Brown Notebook: indicates presence on Catalina Island, March 1973-Sep.1973.
1975	Honolulu, Hawaii working on "rock electricity."
1976-77	Sunnyvale, CA and University of Calif/Berkely: rock electricity work.
1979(?)	Consultant at Stanford Research Institute.
1980	University of North Carolina at Chapel Hill, Project Coordinator for XERXES Project.
1983	Project XERXES continued at California State University (Los Angeles); Brown residing at Avalon, California, as his health was more troublesome.
1985	Brown died at Avalon.

PUBLICATIONS – Reference books and reports on Electrogravitics

Electrogravitics Systems: A New Propulsion Methodology – Volume I by Thomas Valone, PhD, PE. Articles on electrogravitics with diagrams and references, the Aviation Studies Ltd. Reports on the Gravitics Situation. Bonus includes Negative Mass article by Dr. Banesh Hoffman . #611 - 145 pg. book $15

Electrogravitics II: Validating Reports on a New Propulsion Methodology by Thomas Valone. Summary article showing how the Electrokinetic Equation explains pulsed electrokinetics, plus articles on asymmetric capacitors, "How I Control Gravity" by T.T. Brown, and Project Winterhaven. #615 – 150 pg. book $15

The Townsend Brown Electro-Gravity Device. A comprehensive evaluation by the Office of Naval Research, with accompanying documents, issued Sept. 15, 1952. Every page was marked "confidential" until cancelled by ONR. #612 - 22 pg. $10

Thomas Townsend Brown: Bahnson Lab 1958-1960. This rare find is a documentary revealing short segments of all Brown-Bahnson lab experiments. #606- 1 hr silent DVD $20

SubQuantum Kinetics: The Alchemy of Creation. by Dr. Paul LaViolette. Predicts electrogravitic forces. #605 268 pg. book $25

Observations of a Massive Torque Pendulum: Gravity Measurements During an Eclipse by Erwin Saxl and Mildred Allen, Well documented experiment with articles #702 - 52 pages $10

The Zinsser Effect by Thomas Valone, Complete reports on a German pulsed electrogravitic invention and analysis with photos, graphs, and patent, mostly in English. #701- 130 pages $20

Electrogravitics Reference List by Robert Stirniman. Extensive listing of papers, abstracts, websites, and book title/descriptions. Good aid for research in the field. #616 - 55 pages $15

ORDERING INFORMATION: Shipping cost is $5 for U.S. residents. Add $10 for Canada and $15 for overseas. Integrity Research Institute, 5020 Sunnyside Avenue, Suite 209, Beltsville MD 20705. Phone: 800-295-7674 or 301-220-0440. Visit **www.IntegrityResearchInstitute.org** to order online.